Whey - Biological Properties and Alternative Uses

Edited by Isabel Gigli

Published in London, United Kingdom

IntechOpen

Supporting open minds since 2005

Whey - Biological Properties and Alternative Uses
http://dx.doi.org/10.5772/intechopen.79276
Edited by Isabel Gigli

Contributors
Leandro Freire Dos Santos, Giovana Pinto, Letícia Caragnato, Isis Araújo, Andressa Mascarello, André Luís Lopes da Silva, Juan Jauregui-Rincon, Eva María Salinas-Miralles, Norma Angelica Chávez-Vela, Mariela Jimenez-Vargas, Lisbeth Goddik, Paul Hughes, Derrick Risner, Isabel Gigli, Mario Calafat

First published in London, United Kingdom, 2019 by IntechOpen
IntechOpen is the global imprint of INTECHOPEN LIMITED, registered in England and Wales, registration number: 11086078, The Shard, 25th floor, 32 London Bridge Street
London, SE19SG - United Kingdom
Printed in Croatia

British Library Cataloguing-in-Publication Data
A catalogue record for this book is available from the British Library

Additional hard and PDF copies can be obtained from orders@intechopen.com

Whey - Biological Properties and Alternative Uses
Edited by Isabel Gigli
p. cm.
Print ISBN 978-1-83880-925-6
Online ISBN 978-1-83880-926-3
eBook (PDF) ISBN 978-1-83880-927-0

Meet the editor

Isabel Gigli has as a degree in Veterinary Medicine and is a Doctor of Science. After her doctoral research training at the Institute of Biology and Experimental Medicine in Buenos Aires, Argentina, she completed a three-year postdoctoral training at the University of Cornell, USA. In addition, she worked for two years at the University of Palermo, Italy. Currently, Dr Gigli is a full-time professor at the School of Agriculture at the National University of La Pampa, Argentina, sharing her passion for teaching with her interest in scientific writing.

Contents

Preface

The Food and Agriculture Organization of the United Nations has estimated that by 2050, food production should increase by 70% to feed the growing population. If we continue to produce the way we do, environmental pollution will undoubtedly be greater. It seems a perverse contradiction that while today there are more than 800 million undernourished or malnourished people in the world, a third of all the food produced is lost or wasted during the production process. This is by no means the cause of food injustice, but is certainly part of the problem. It is urgent that we rethink the way we are producing food, and especially our use of agro-industrial by-products. Whey - a cheese by-product - contains 50% nutrients and 90% water of original milk. However, whey is discarded as waste in most small and medium cheese industries throughout the world. The available technologies to process dairy industry by-products are expensive in terms of equipment acquisition and energy cost. This results in a dilemma: is it more profitable to discard whey as an effluent instead of elaborating it with added value. *Whey - Biological Properties and Alternative Uses* proposes to break this paradigm and has invited researches committed to this problem to present their proposals and results for the use of dairy by-products beyond the traditional ones, such as direct use for animals, dehydration, or protein concentrate. This book is an excellent opportunity for graduate students and researchers to become aware of the problem of the misuse of agro-industrial by-products. Reducing food waste is everyone's responsibility. We must stop thinking about agro-industrial products as waste and find production alternatives.

Dr. Isabel Gigli
Agriculture College,
National University of La Pampa,
La Pampa, Argentina

Section 1

Whey

Chapter 1

Introductory Chapter: Dairy By-Products - Why Should We Care?

Isabel Gigli and Mario Calafat

1. Introduction

This book focuses on low investment alternative use of dairy by-products. Whey has application in both the pharmacology and nutritional industry. However, three main problems affront cheese makers when they try to process whey: the short half-life, the cost of refrigerating, and transportation cost. All these make the use of whey economically difficult for small and medium manufacturers. Therefore, in most cases, whey ends up as agro-industrial waste. This represents a loss of valuable opportunities and also, as explained below, represents a high environmental impact. In the different chapters, the authors offer alternative biotechnology processes (**Figure 1**). The ultimate goal of the book is to break the paradigm of considering milk by-product as a waste. This introductory chapter provides the global context in which the book was conceived: starts with a historical and current perspective of the consumption of dairy products, continues with the composition of the by-products, followed by our experience of using whey as culture media to produce mineral organic supplement and then isotonic lactose-free beverage, and closes with a general conclusion.

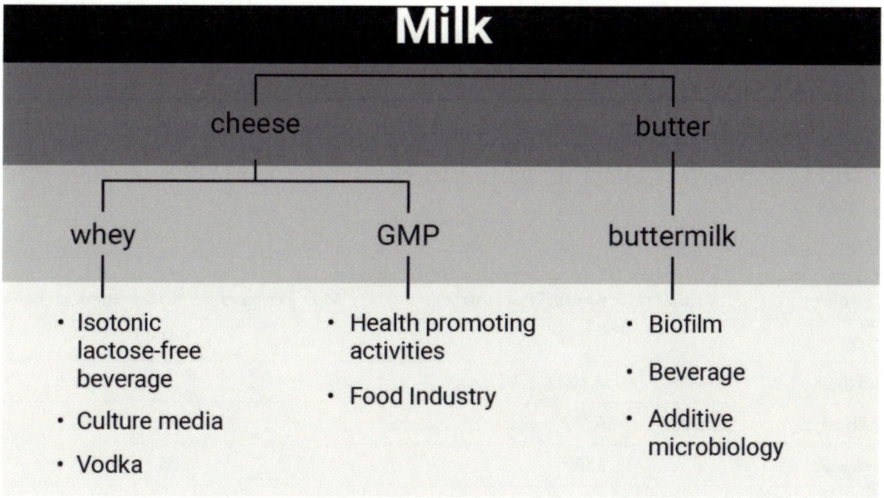

Figure 1.
Alternative use of whey and buttermilk discussed in this book.

2. Dairy products consumption: past and present

Cow milk has been part of the human diet for 11,000 years. In the 1970s, archeologist Peter Bogucki while excavating a Stone Age site in Poland found a leaky clay pot. It was not until 2011 when molecular studies performed by Mélanie Roffet-Salque identified milk fat residues [1]. The researcher, a UK geochemistry, concluded that the container constituted evidence that prehistoric farmers used the ceramic as sieves to separate milk solids from whey. This pot is the oldest known evidence of cheese making in the world. Cheese has been a nutritional contribution in the diet of many cultures. Nowadays, 6 billion people around the world consume milk and milk products [2]. In South America alone, milk production reached 61.8 million tons in 2017 [3]. In this region, the largest producers are Brazil, Colombia, and Peru. In Brazil, fluid milk market predominates over manufacturing dairy products. While in Argentina, Mexico, Chile, and Colombia, the cheese and butter markets are more important than fluid milk [4] as it is shown in **Table 1**.

The need to produce safe and nutritious food without environmental impact is a global challenge. Technology has contributed to improve agricultural and livestock productivity, but at the same time, the increase in production has had a negative impact, such as environmental and water pollution, deforestation, and biodiversity loss. The Food and Agriculture Organization of the United Nations estimates that by 2050, food production should increase by 70% over current production [5]. This information together with the information that currently a third of the global food is lost or waste [5] highlights the importance of rethinking food production. The term **food loss** refers to losses that occur during the supply chain between the producer and the market (e.g., during sowing, harvesting, or transporting). **Food waste** refers to the nonuse or nonfood use that can be given to raw material safe and nutritious suitable to be converted into food. The last one is the case of dairy by-products, especially in developing countries where the energy cost for technological processes such as drying or protein purification makes it economically difficult. While we are being inefficient in food production, around 870 million people do not have access to sufficient dietary energy and as a consequence suffer chronic malnutrition. Therefore, nothing should justify the voluntary loss of raw materials that could be transformed into food. The nonuse of by-products that could be transformed into food represents a waste of food and also a waste of the resources used to produce them. This inefficiency also harms environmental sustainability. Lactose is a strong pollutant due to its high oxygen demand. It is important to find alternative uses of milk by-product to avoid or at least reduce food waste.

Country	Annual milk production ($\times 10^6$)	Milk percentage (%) destined to cheese
Argentina	11.338	41
Brazil	33.400	38
Colombia	6.772	35*
Uruguay	2.100	35

Table 1.
Annual milk and cheese production in South America (2012–2013) [4].

3. Whey

Whey is the remaining liquid that is produced after milk has been curdled and strained during the manufacturing of hard and semihard cheese. Depending on the process of casein precipitation, whey can be acid or sweet. Acid whey is formed when lactic acid bacteria are added to milk, and sweet whey is formed when the coagulation process is started by adding chymosin. Whey contains 90% of milk water and 50% of milk nutrients (**Table 2**). The major whey proteins are lactoglobulin, lactalbumin, serum albumin, immunoglobulins, and glycomacropeptide, while minor whey proteins include lactoperoxidase and lactoferrin among other proteins (**Figure 2**). These globular proteins are water soluble and contain all nine essential amino acids. **Tables 2** and **3** show the composition and the vitamins present in whey, respectively. As it can see, whey is nutritional and, in spite of that, it is rarely used as such. One problem of using whey as a food is the high concentration of lactose that makes it difficult to digest especially for people intolerant to this carbohydrate (lactose intolerance). For lack of alternative uses, what happens is that it ends up being discarded as an effluent.

The amount of discarded whey is difficult to quantify for obvious reason. The Food and Agriculture Organization (FAO) estimated that 40 × 10^6 l/day of whey are produced in South America, and the largest amount is discarded. In Argentina, the dairy industry produces around 11,338 × 10^6 l/year. Approximately, 41% is used for the production of cheese. This volume is equivalent to producing 4.015 × 10^6 l/year of whey. Fifty-five percent of that volume is used in part to feed animals and most of it is discarded as waste [4].

Whey has a negative environmental impact for both soils (due to mineral concentration) and water (due to lactose concentration) Although this is well known, what is observed in most cheese factories is that whey is discarded, without treatment or with a minimum treatment that fails to reduce the chemical oxygen demand (COD) to acceptable values (marcos murcia, personal communication). As a way to contribute to the search of solutions to this problem, our group seeks simple and economical alternatives to implement the use of whey. Below, two alternatives are discussed. First, we discuss the use of whey as a culture medium for yeast enriched in specific minerals and, second, a beverage elaboration carried out by a group of students from an agricultural school under our direction. This last experience, we believe, is important to visualize the problem and promote the local production.

3.1 Whey as a culture media to obtain selenium-enriched yeast

The concentration of minerals present in the soil varies according to the geographical area. When a mineral is deficient in soils, livestock needs to be

Composition	%
Fat	0.7
Total protein	0.9
Lactose	4.5
Total solid	6.7
No fat solid	6.1

Table 2.
Whey composition [6].

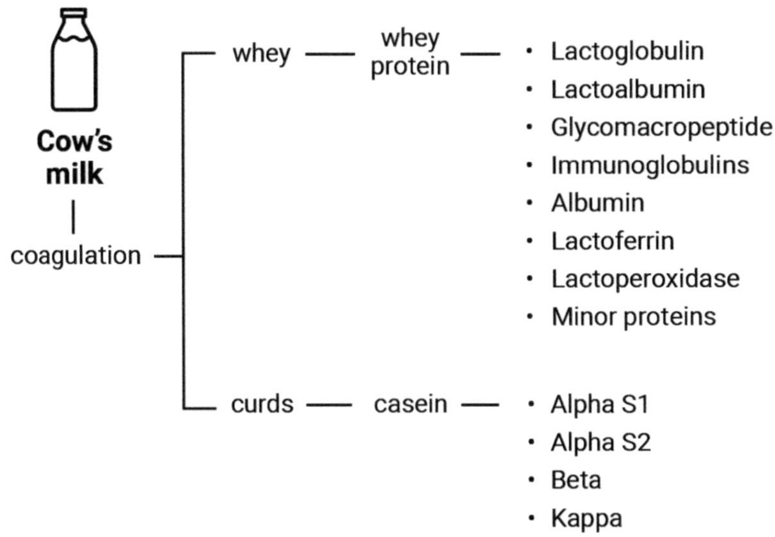

Figure 2.
Whey and curd protein composition.

Amino acid	%
Thiamine	0.3
Folic acid	0.7
Niacin	1.2
Riboflavin	0.2
Choline	108.0
Pantothenic acid	4.0

Table 3.
Vitamin concentration in whey [7].

supplemented. In the semiarid region, for example, a low concentration of selenium is observed [8]. Selenium is a cofactor of different enzymes such as glutathione peroxidase, which is involved in the antioxidant systems [9]. Also, selenium plays an important role in thyroid hormone synthesis [10]. Although the animal requirement for this mineral is low (300 μg/ dry matter) [11], selenium deficiency affects health and productivity. In cows, immunological suppression, placental retention, and decreased milk production have been reported associated with selenium deficient [12, 13]. In calves, diets deficient in selenium can cause a lethal condition called white muscle [14]. In order to find an alternative use to whey and at the same time offer a solution to the geographical areas that have selenium deficiency, we developed an economical culture using whey for the growth of yeast expressing lactase (Kluyveromyces DSM 11954). In this way, lactose is used to obtain selenium-enriched biomass. The whey-based culture medium (whey, $(NH_4)_2SO_4$ and K_2PO_4) was supplemented with 20 μg/ml of Na_2SeO_3. At the end of the process, a total of 550 g wet cell weight (WCW) was equivalent to 85 mg/kg yeast selenium concentration. We then studied the effect of selenium-enriched yeast on calves. Six calves received 7 g of K. *marxianus* daily for 10 days (0.60 mg of selenium/animal/day). The supplement was offered individually to the animals, mixed in a small amount of grains. The animals showed no signs of rejection of the supplement. The level of selenium in the blood was measured at Day 0 (before supplementation) and at Day 10 of treatment. The animals showed a significant increase in selenium blood ($p < 0.005$) (**Figure 3**) [6]. The use of whey as a culture medium is an inexpensive way to produce organic mineral supplements and at the same time reduces the environmental impact caused by the concentration of lactose in wasted milk by-product.

3.2 Whey as an isotonic, lactose-free beverage

Water footprint is the amount of direct and indirect water used to produce a product. It is estimated that a glass of milk (200 ml) requires 250 l of water (1 water l/1000 milk l) [15]. Taking into account that whey contains 90% milk water and is discarded without use, it is surprising that no further efforts are made to avoid the waste of such a volume of water. In addition to water, whey—as mentioned above—is a source of all essential amino acids. In Chapter 2 of this book, a fermented beverage process from whey is described. Here, whey is discussed as an isotonic drink. Whey has been promoted to be used as a sports drink as it contains all minerals to replace the electrolyte losses in sweat and carbohydrate [16, 17]. **Table 4** shows whey and commercial sport drinks mineral concentration. Isotonic drinks help prevent blood sodium dilution, a dangerous situation that occurs when athletes, especially in long distance events, drink water in excess [21]. Athletes need to replace water and minerals during and after an endurance event. Mineral losses by sweat include sodium, potassium, magnesium, chloride, and calcium. All of these minerals play important biochemical and physiological roles in the body. Another nutritional advantage of whey protein is that it has a high level of leucine (11.8% of total protein), which is important in sports supplementation when the objectives are muscle repair and growth. For example, Hamarsland et al. [22] reported higher blood leucine concentration and higher muscle protein rates after exercise when athletes consumed native whey, compared to milk or WPC-80. Reitelseder et al. [23] quantified labeled L-[1-13C] leucine in muscle and compared two different treatments: casein and whey ingestion. The authors observed no difference in protein muscle synthesis. These results are important because they valorize the use of whey compared to more costly technological processes. The point that must be emphasized is that the promotion of whey as a beverage will help reduce environmental pollution and develop a new commercial option to use a by-product that, otherwise, is considered a waste.

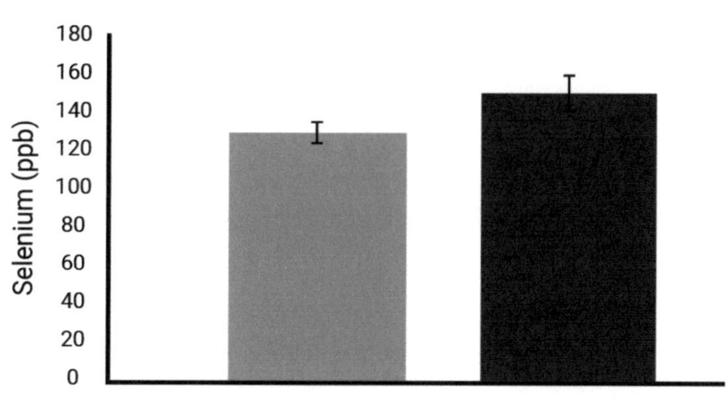

Figure 3.
Selenium blood levels (ppb) in calves before (Day 0) and the day after selenium-enriched yeast supplementation (Day 10).

Mineral	Sweet whey (mg/100 ml) [18]	Acid whey (mg/100 ml) [18]	Gatorade (mg/100 ml) [19]	Powerade (mg/100 ml) [20]
Calcium	92.8	36.5	—	—
Phosphorus	58	43	—	—
Magnesium	9	6.5	—	—
Potassium	58	43	45.8	104.2
Sodium	39.5	45.5	12.5	24.2

Concentration in whey can vary according to whey processing.

Table 4.
Mineral concentration in sweet whey, acid whey, and commercial sport drink Gatorade and Powerade.

Keeping the idea of finding alternative products for whey, the authors of this chapter collaborated with students of Agriculture School (Escuela Agrotécnica Victorica, La Pampa, Argentina) to produce a lactose-free beverage based on whey. The students developed the product as part of their laboratory assignment for the Nutrition course. **Figure 4** shows the production flow, and **Figure 5** shows the final product. In this way, the students visualized and understood that with a simple process, what is considered a waste can be transformed into food.

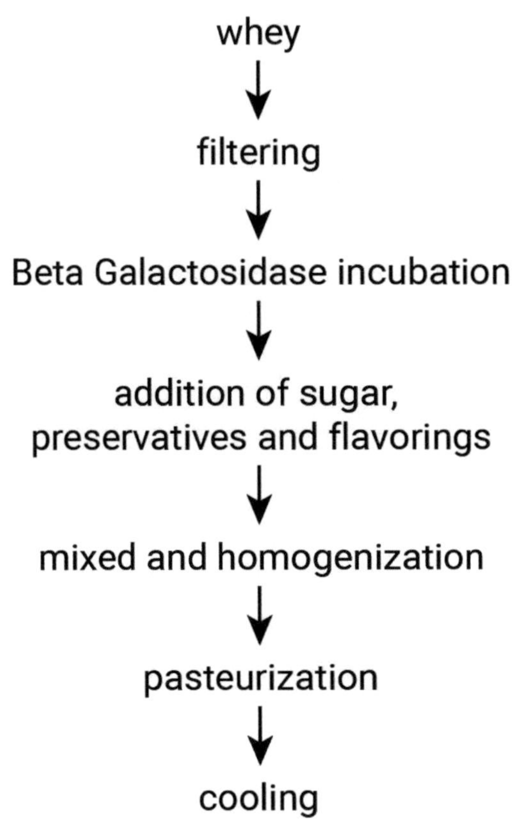

Figure 4.
Lactose-free whey beverage production flow.

Figure 5.
Lactose-free whey beverage produced by students of Agriculture School Victoria (Escuela Agrotécnica Victorica, La Pampa, Argentina).

4. Conclusion

Whey is still considered a by-product with no economical value despite its biological properties and despite all the alternative uses that have been proposed. The uses of whey as a culture medium or as a beverage, as discussed here, are examples of how simple processes can contribute to the development of new products and at the same time to reduce environmental impact. We are intensifying production systems and genetically improving livestock to produce more milk; meanwhile, we are discarding 50% of the nutrients in milk as whey. This should be considered also as part of lack of animal welfare. We do not need to produce super cows, with all the metabolic stress that it causes to the animal; we need to rethink the systems of food elaboration.

It is important to discuss the way we are producing food and the biological value of the food we want to consume. Environmental and production policies that favor the use of by-products are required. We need to break the paradigm of considering milk products as waste and develop new strategies for use.

Acknowledgements

The authors want to express their gratitude to Ing. Agr. Claudia Gorozurreta (Agrotechnical School "Florencio Peirone" Victorica) for her work with the students during the elaboration of whey beverage; to Graphic Designer Micaela Maizon for the figures; and Lic. Ana Paez for her thoughtful revision of English.

Author details

Isabel Gigli* and Mario Calafat
Agriculture College, National University of La Pampa (UNLPam), La Pampa, Argentina

*Address all correspondence to: igigli@agro.unlpam.edu.ar

References

[1] Salque M, Boguck PI, Pyze J, Sobkowiak-Tabaka I, Grygiel R, Szmyt M, et al. Earliest evidence for cheese making in the sixth millennium BC in northern Europe. Nature International Journal of Science.2013;**493**:522-525

[2] Food and Agriculture Organization of the United Nations. Chapter 7: Dairy and dairy products. OECD-FAO Agricultural Outlook 2018-2027. 2018. Available from: http://www.agri-outlook.org/commodities/Agricultural-Outlook-2018-Dairy.pdf

[3] Food and Agriculture Organization of the United Nations. Dairy Market Review. Rome: FAO; 2018. Available from: http://www.fao.org/3/CA2857EN/ca2857en.pdf

[4] Muset G, Castells ML, editors. Valorización del lactosuero. San Martín: Instituto Nacional de Tecnología Industrial—INTI. 2017

[5] Food and Agriculture Organization of the United Nations. 2018. Available form: http://www.fao.org/fileadmin/templates/wsfs/docs/expert_paper/How_to_Feed_the_World_in_2050.pdf [Online]

[6] Gurdo N, Calafat M, Noseda DG, Gigli I. Production of selenium-enriched yeast (*Kluyveromyces marxianus*) biomass in a whey-based culture medium. American Journal of Biochemistry and Biotechnology. 2018;**14**(2):175-182

[7] McDonough FE, Hargrove RE, Mattingly WA, Posati LP, Alford JA. Composition and properties of whey protein concentrates from ultrafiltration. Journal of Dairy Science. 1974;**12**:1438-1443. s.l.: 57

[8] Gil S, Hevia S, Dallorso M, Resnizk S. Selenium in bovine plasma, soil and forage measured by neutron activation analysis. Arquivo Brasileiro de Medicina Veterinária e Zootecnia. 2004;**56**:264-266

[9] Smith P, Tappel A, Chow C. Glutathione peroxidase activity as a function of dietary selenomethionine. Nature. 1974;**5440**:392-393. s.l. 247

[10] Beckett GF, Nicol P, Rae S, Beech Y, Guo JA. Effects of combined iodine and selenium deficiency on thyroid hormone metabolism in rats. The American Journal of Clinical Nutrition. 1993;**57**(2):240s-243s

[11] National Research Council. Chapter 6: Minerals. In: Nutrient Requirements of Dairy Cattle. 7th revised ed. Washington, DC: The National Academies Press; 2001. DOI: 10.17226/9825

[12] Julien WE, Conrad HJ, Moxon JA. Selenium and vitamin E and incidence of retained placenta in parturient dairy cows. Journal of Diary Science. 1976;**11**:1954-1959

[13] Moeini NM, Karami H, Mikaeili E. Effect of selenium and vitamin E supplementation during the late pregnancy on reproductive indices and milk production in heifersv. Animal Reproduction Science. 2009;**1-3**:109-114

[14] Andrews E, Hartley W, Grant A. Selenium responsive diseases of animals in New Zealand New Zealand Veterinary Journal 1968;**16**:3-17

[15] Mekonnen MM,Hoekstra AY. The green, blue and grey water footprint of farm animals and animal products . Value of water research report Series 48, 2010 UNESCO-IHE

[16] Li L, Sun FH, Huang WY, Wong SH. Effects of whey protein in carbohydrate-electrolyte drinks on postexercise rehydration. European Journal of Sport

Science. 2018;**18**(5):685-694. DOI: 10.1080/17461391.2018.1442499 [Epub Mar 1, 2018]

[17] Hulmi JJ, Laakso M, Mero AA, Häkkinen K, Ahtiainen JP, Peltonen H. The effects of whey protein with or without carbohydrates on resistance training adaptations. International Society of Sports Nutrition. 2015;**12**:48. DOI: 10.1186/s12970-015-0109-4 (eCollection 2015)

[18] Wong NPD, LaCroix E, McDonough FE. Minerals in whey. Journal of Dairy Science. 1978;**61**:1700-1703

[19] Gatorade Canada. [Online] April 17, 2019. Available from: http://www.pepsico.ca/en/Brands/Gatorade.html

[20] Powerade products. [Online] April 17, 2019. Available from: https://www.powerade.com/products/fruit-punch/

[21] Noakes TD, Sharwood K, Speedy D, Hew T, Reid S, Dugas J, Almond C, Wharam P, Weschler L. Three independent biological mechanisms cause exercise-associated hyponatremia: Evidence from 2,135 weighed competitive athletic performances. 2015 Proceedings of the National Academy of Sciences of the United States of America. 2015;**102**(51):18550-18555 [Epub Dec 12, 2005]

[22] Hamarsland H, Nordengen AL, Nyvik Aas S, Holte K, Garthe I, Paulsen G, et al. Native whey protein with high levels of leucine results in similar post-exercise muscular anabolic responses as regular whey protein: A randomized controlled trial. Journal of the International Society of Sports Nutrition. 2017;**21**(14):23

[23] Reitelseder S, Agergaard J, Doessing S, Helmark IC, Lund P, Kristensen NB, et al. Whey and casein labeled with L-[1-13C]leucine and muscle protein synthesis: Effect of resistance exercise and protein ingestion. American Journal of Physiology-Endocrinology and Metabolism. 2011;**300**(1):E231-E242. DOI: 10.1152/ajpendo.00513.2010. [Epub Nov 2, 2010]

Chapter 2

Whey to Vodka

Paul Hughes, Derrick Risner and Lisbeth Meunier Goddik

Abstract

Whey production can be an economic and environmental problem for small creameries and acid whey producers. The fermentation and distillation of whey not only eliminates the cost of disposing whey as waste while minimizing environmental impact but adds a revenue option through production of a value-added product. *Kluyveromyces marxianus* is typically utilized to ferment the pasteurized and pretreated whey. The fermented product contains approximately 3% ethanol v/v. Various options for distilling may be utilized such as a simple two-pot system or a more complex four-stage system to assure production of a neutral spirit. Quality of the distilled spirit is impacted by whey source, whey pretreatment, fermentation conditions, and the distilling process.

Keywords: *Kluyveromyces marxianus*, fermentation, distillation, spirits, ethanol, still, Carbery method

1. Introduction

Whey processing is a mature manufacturing sector. More than 75 years have passed since multiple effect evaporators and spray dryers were developed and applied to whey processing [1]. Nevertheless, the technology continues to evolve. The initial processes focused on removing water and concentrating all solids-non-fat into dry powders. Today, membranes, ion exchange resins, and chromatography are some of the new unit operations routinely applied in the processing of an increasingly diverse assortment of powders originating from whey.

This development has greatly benefitted larger cheese producers as these powders generally provide significant revenue potential. Unfortunately, smaller scale cheese processors are rarely able to benefit from these products. Whey powder facilities are expensive to construct and are therefore not an option for smaller cheese companies.

Large-scale cheese makers in the US typically only produce one type of cheese such as cheddar or mozzarella. This leads to production of large volumes of sweet whey streams with consistent composition that are well suited for current whey manufacturing facilities. In contrast, smaller specialty cheese producers tend to produce multiple different cheese types and must deal with different whey streams. While most hard renneted cheeses produce relatively similar whey streams, the lactic cheeses such as cottage or cream cheese along with Greek yogurt create acid whey. Acid whey primarily differs from sweet whey in mineral and acid content. Specifically, acid whey may have twice the calcium content and more than 10 times the lactic acid content as compared to sweet whey. The high levels of lactic acid interfere with the drying process as it contributes to forming sticky powder agglomerates within dryers. Consequently, acid whey cannot be easily processed into whey powders.

Giving these limitations, small-scale cheese processors and acid whey producers have limited options for whey disposal. At best, they aim to dispose of whey without a cost. This could involve using whey as an animal feed source, land application, or disposal in farm lagoons. All of these options have potential negative consequences. Dragone et al. suggested that 47% of whey produced in Portugal (mostly from small scale producers) was disposed through land application or directly into streams [2]. The environmental consequences of this can be significant due to the high Biological Oxygen Demand (BOD) and Chemical Oxygen Demand (COD) of whey, which are 40–60 and 50–80 g/L respectively [3]. This leads to depletion of dissolved oxygen when disposed into lakes and streams. Whey does not appear to negatively impact the flavor of beef from cattle fed whey [4, 5], although some negative impacts such as acidosis and diminished carcass grade have been noted [6]. In addition, feeding whey back to the livestock at a farmstead creamery will likely increase the risk of phage development.

Nevertheless, self-disposal may be favorable compared to paying for disposal through municipal wastewater treatment systems or paying others to haul the whey away for disposal. Rates for disposal of waste through municipal water/waste treatment rates are based upon the mass of BOD being removed at the treatment facility thereby making it an expensive waste treatment option for whey. In fact, it may not even be an option as some municipalities refuse to treat whey. A recent (2015) unpublished survey of specialty cheesemakers in the US revealed that most of the very small artisan cheese makers manage to dispose of whey at no cost through feeding to own or local neighborhood animals. However, as soon as cheese production increases above 5000 kg/year, most are obliged to pay for disposal at rates up to $105/1000 kg of whey. This demonstrates that whey disposal can be a significant expense for medium scale cheese processors that are too small to produce whey powders and too large to dispose of whey through feeding or other small-scale disposal. As profit margins for small-scale cheese makers are tight [7], whey disposal costs can significantly impact business sustainability.

Due to these challenges, small-scale whey producers are continuously looking for whey disposal options. The fermentation and distillation of whey can be done to produce bioethanol or a potable spirit. The fermentation and distillation of whey to produce potable spirits may be a potential value-added option for small scale cheese makers. Not only does this allow for concentrating the initial whey stream, but it also enables the production of an additional high-priced product. For example, if a 750 ml bottle of vodka sells for $30 that would translate to approximately $1–1.5 per L of initial whey. This could potentially create as much revenue as the corresponding cheese.

2. Commercial whey spirits

The concept of producing whey-based spirits is not new. This process has been explored scientifically since the 1940s and the Carbery process was developed and commercialized in 1978 to produce potable ethanol from whey on an industrial scale [8, 9]. Analysis has been conducted illustrating that whey based spirits are composed of volatile compounds similar to other spirits and are safe for consumption [2]. Currently in New Zealand, potable ethanol is being produced using the Carbery process and is exported to Asian markets [9, 10]. There are multiple examples worldwide of commercially available whey-based spirits. All of these products highlight the dairy/whey connection; both on the label and in product description that emphasizes creamy flavors. They are all marketed as premium products and

Figure 1.
Commercially available whey-based spirits: Bertha's revenge gin, black cow vodka, Vermont white vodka, and sheep whey vodka.

sold at high prices. This demonstrates that consumers appreciate distilled spirits produced from dairy sources. Below is a summary of four commercial whey-based spirits (pictured in **Figure 1**).

- Bertha's Revenge and Slough Bertha are produced at Ballyvolane Guesthouse in Ireland (https://ballyvolanespirits.ie). The product is named after Bertha, a Droimeann cow from Sneem in Co. Kerry, who apparently lived to be 49 years old. Although this gin is labeled as an Irish milk gin, it is produced from fermented sweet whey. The alcoholic whey is distilled three times and flavored with local botanicals.

- Black cow vodka is produced in England (https://www.blackcow.co.uk). This vodka sells for a premium price. The product has significant worldwide distribution and in deference to regulations in various countries is sold in select countries as a spirit instead of vodka to recognize that it is not based on grains or potatoes.

- An American version is Vermont white vodka from Vermont Spirits (http://www.vermontspirits.com). Vermont Spirits converts multiple local agricultural products to spirits. Tasting notes for this product describes it as: "a traditional vodka with a bracing yet moderately light medicinal approach, then a finish that fades into a nice and lingering sweetness. Creamy, with just a hint of bittersweet chocolate."

- Sheep Whey Vodka is produced at a Tasmanian artisan creamery (http://grandvewe.com.au). This product is the only one of the four spirits that is produced at the creamery. In appreciation of the creativity of this product, it won Champion Vodka of Australia at the World Vodka Awards 2017 in London, along with the 2017 award for Australian Beverage of the year.

3. Controlling the whey source

An important conclusion from a recent study by Risner et al. is that aroma compounds within spirits differ significantly based on whey source [11]. Therefore, it is essential to understand and control the whey source prior to starting commercial fermentation and distillation of whey, composition of sweet whey depends on a wide variety of factors. Within cheese types, milk pretreatment and cheese processing parameters such as filtration, pasteurization, starter, rennet, and salting will all impact whey composition [12, 13]. Among different cheese types, the parameters listed above have even more impact with the largest differences associated with lactic curd cheeses such as cottage cheese. In addition, external factors such as feed, season, and lactation influence whey composition [14]. This is particularly important for whey from goat and sheep milk cheeses as these animals are often on seasonal lactation schedules [15]. Small variations in compositions likely do not greatly affect fermentation and distillation; nevertheless, this can be a concern when striving to produce a consistent product.

4. Whey pretreatment

Although whey from different sources vary, there are tools available to pretreat the whey prior to fermentation and distillation. Traditionally, whey clarifiers are used to remove casein fines while whey separators are used to remove whey cream. This leaves behind non-fermentable substrates such as whey proteins, minerals, and acids, which do not contribute to the production of distilled beverages. Although whey proteins are soluble, they may precipitate when exposed to heat during pasteurization or during distillation, which could interfere with operation of the still. Therefore, some method of protein removal, such as ultrafiltration, would be beneficial prior to fermentation. Removal of other potentially interfering compounds such as minerals and acids could be achieved through nanofiltration. Nanofiltration has the additional advantage of concentrating lactose to increase the concentration of fermentable substrate within whey, which would essentially improve fermentation and distillation efficiencies. It is important to note that these unit operations are expensive and resource intensive and therefore not likely to be used in artisan dairy processing. Nevertheless, membrane units are utilized in some specialty cheese facilities and could therefore be a relevant option.

5. Whey to commercial spirit

The Carbery method is the industrial method used to convert whey/whey permeate to ethanol [8, 9]. The method is similar to other industrial ethanol production processes in that a microbial fermentation is performed to convert sugars within a substrate to ethanol and an extractive distillation occurs to concentrate and separate the ethanol from other volatile compounds. Once distillation has occurred the spirit can be treated as any other distilled spirit for subsequent processing (**Figure 2**).

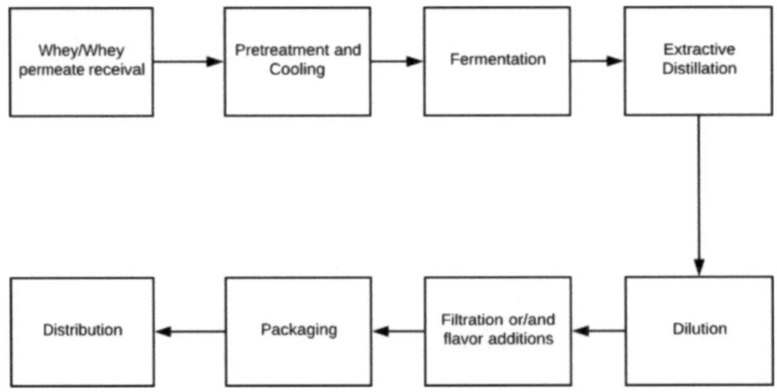

Figure 2.
Process overview of spirit production via the Carbery method.

There are several key differences in the Carbery method when compared to traditional spirit production. Whey/whey permeate is readily fermentable and a sugar conversion step such as mashing or cooking is not necessary. Whey/whey permeate should arrive at the facility well above the optimum fermentation temperature and must be cooled before inoculation. The main fermentable sugar within whey is lactose, which cannot be utilized by *Saccharomyces cerevisiae (S. cerevisiae)*, the yeast generally used for ethanol production. *Kluyveromyces marxianus (K. marxianus)*, a lactose fermenting yeast is used to convert lactose to ethanol. The lactose levels within raw whey only allow for the production of a "beer" or "wash" with ethanol concentrations of 2–3% v/v. Whey permeate may be concentrated but ethanol production is limited by the sensitivity of *K. marxianus* to increased solute concentrations and ethanol. This low concentration of ethanol will increase the energy requirements during the distillation process. A beer still, extractive distillation unit and a rectifier are used during the extractive distillation process. A demethylizer is not employed during the extractive distillation process [8, 9] as very little methanol is formed during the fermentation process. The dilution, filtration, flavor additions, packaging, and distribution occur in a manner comparable to other spirits produced in a traditional manner.

6. Conversion of lactose to ethanol

Lactose [O-β-D-galactopyranosyl-(1→4)-D-glucopyranose] is a reducing sugar and disaccharide composed of β-1,4 glycosidically bonded galactose and glucose residues. Lactose is the primary carbohydrate constituent of whey and whey permeate [16, 17]. The conversion of lactose to ethanol is a two-step process. First, lactose must be hydrolyzed to galactose and glucose and then alcoholic fermentation occurs to produce ethanol.

6.1 Methods of lactose hydrolysis

The enzymatic hydrolysis of lactose is the most common method of lactose hydrolysis (**Figure 3**) and can be achieved in several ways. The common industrial conversion of lactose to ethanol uses an ethanol producing microbe, *K. marxianus* which enzymatically hydrolyzes lactose [8, 9]. Whey or whey permeate is cooled to the microbe's optimum fermentation temperature and then inoculated. Hydrolysis of lactose is achieved intracellularly via β-galactosidase and the organism subsequently metabolizes the constituents to produce ethanol [18]. It should be noted that the traditional brewing and distilling yeast used to produce ethanol, *S. cerevisiae* does not express the genes necessary to produce β-galactosidase, as an alternative a β-galactosidase producing yeast, *K. marxianus* is used. Genetic engineering of *S. cerevisiae* to produce β-galactosidase has been explored on an experimental scale

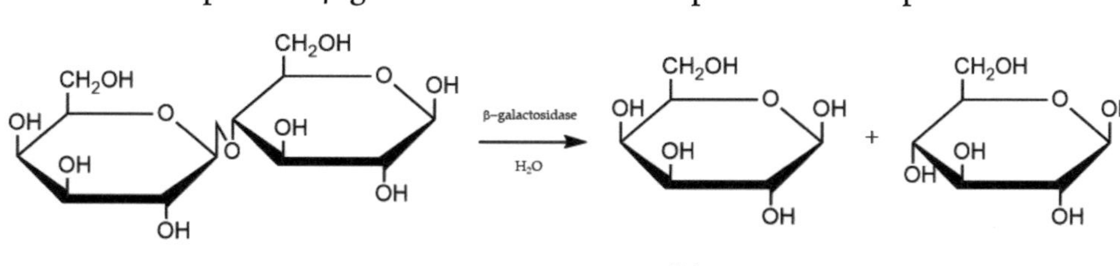

Figure 3.
Enzymatic hydrolysis of lactose.

for bioethanol production, but to the authors' knowledge this is not being used in beverage production [19–21].

A common method of lactose hydrolysis in dairy product production is the addition of lactase, an exogenous enzyme belonging to the β-galactosidases family [22–24]. The addition of this enzyme requires no additional processing equipment and lactase is widely available. Using lactase to hydrolyze lactose allows for the use of microbes, which do not produce β-galactosidase, to be used in the fermentation. This approach has been explored and documented on an experimental scale for bioethanol production [25, 26].

Other methods of hydrolysis of lactose include the use of immobilized enzyme systems, membrane reactor processes used to recover enzymes/cells and acid hydrolysis [24, 27]. Immobilized enzyme and membrane reactor systems could help reduce cost because both are enzyme conservation processes, but they require additional processing technology and are not widely implemented commercially. Acid hydrolysis requires the use of ultrafiltration because the whey permeate stream must be free of protein. The process involves the acidification and short heat treatment ranging from approximately 100–150°C. This treatment causes a brown discoloration in serum which requires color removal and purification steps [24, 27]. The color removal process would not be necessary during ethanol production. While these technologies and processes are currently not used in the commercial conversion of whey to ethanol, some have been explored to increase production efficiency [26, 28–30].

6.2 Fermentation after lactose hydrolysis

Alcoholic fermentation is a form of anaerobic energy production commonly used by plants, yeast and other microbes [31]. This metabolic pathway has been exploited by humans for food and beverage production for several millennia. During industrial production of ethanol from whey, an ethanol-fermenting strain of *K. marxianus* is used to convert lactose into ethanol. This strain of *K. marxianus* is used because it can intracellularly hydrolyze lactose and efficiently produce ethanol.

Alcoholic fermentation has two distinct phases. The first phase is glycolysis which converts glucose to pyruvate. The glycolytic pathway is common to nearly all

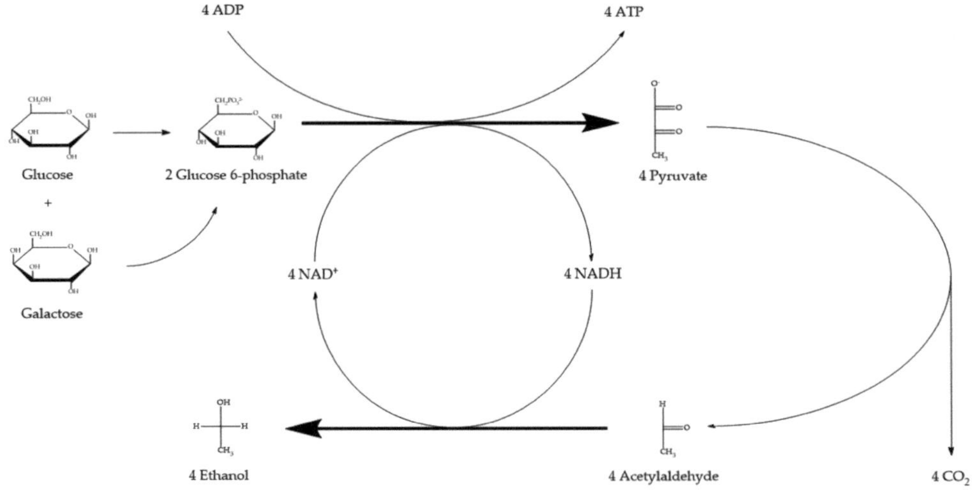

Figure 4.
Alcoholic fermentation after lactose hydrolysis.

cells and generates adenosine triphosphate (ATP) which is used for intracellular energy transfer. Galactose is enzymatically converted to glucose 6-phosphate, an intermediate product of glycolysis (**Figure 4**). The conversion of galactose to glucose 6-phosphate is a four step process; however, the cellular energetic cost is the same as the phosphorylation of glucose. The outcome of this glycolysis process is net production of 4 ATP, the conversion of glucose and galactose to 4 pyruvate molecules and the reduction of NAD^+ to NADH.

The second phase of alcoholic fermentation converts pyruvate into ethanol to regenerate NAD^+ used during glycolysis. Pyruvate is decarboxylated enzymatically which results in the production of CO_2 and the formation of acetaldehyde. The reduction of acetaldehyde to ethanol is catalyzed by alcohol dehydrogenase and NAD^+ is replenished in the process [31]. Ethanol is then passively diffused from the cell into the fermentation substrate.

7. Fermentation organisms

There are few yeast species which assimilate lactose to produce ethanol [32]. *K. marxianus* is the microorganism widely used in industrial lactose to ethanol conversion. Other microorganisms used in industrial food and beverage manufacture have been examined at an experimental scale for their suitability for lactose to bioethanol conversion. These organisms include *K. lactis, S. cerevisiae* and *Escherichia coli.* Use of genetically engineered organisms for alcoholic beverage manufacture is currently not a common commercial practice. This is likely due to perceived consumer concerns about the consumption of genetically modified organisms.

7.1 *K. marxianus* and considerations for lactose to ethanol conversion

K. marxianus ability to convert lactose to ethanol is widely reported in scientific studies concerning bioethanol production. *K. marxianus* is the fermentative organism used for large scale manufacture of potable spirits and bioethanol produced from whey/whey permeate [8, 9]. Scientific studies often reference *K. fragilis* as a lactose-fermenter; however, it is currently synonymous with *K. marxianus* [33]. Several studies have investigated the use of *Candida pseudotropicalis* as the fermentative organism for lactose to ethanol conversion [10]. *C. pseudotropicalis,* also referred to as *Candida kefir*, is the anamorph (asexual reproductive stage) of *K. marxianus* [33]. The species *K. marxianus* have a high degree of genetic variation and each strain's ability to produce ethanol can vary widely [34–36]. This is likely due to the species being present in a wide range of habitats [35]. *K. marxianus* is widely considered to be a Crabtree-negative organism, meaning the organism will preferentially respire instead of ferment when oxygen and glucose are abundant [37]. *K. marxianus* carries the genes necessary for fermentation and strains have been reported as Crabtree-positive (preferentially ferments in presence of oxygen and an abundance of glucose) [37, 38]. The ethanol tolerance of *K. marxianus* is lower than *S. cerevisiae* and can limit ethanol production [39]. Inhibition of ethanol production can occur at ethanol concentrations as low as 45–52 g/l or approximately 5.5–6.5% v/v [40]. Supplementation or concentration lactose within whey or whey permeate can cause substrate inhibition and limit ethanol production. This trait appears to be strain specific with reports varying of ethanol production inhibition at lactose concentration of 108–200 g/l [41, 42]. This wide variation in reported ranges highlights the importance of purchasing the proper fermentative strain of *K. marxianus* to meet each lactose to ethanol producer's requirements.

K. marxianus is generally recognized as safe (designated GRAS), which is advantageous for potable spirit producers because the yeast biomass can be further processed for livestock or human consumption. It has been reported that the fermentation process can reduce the biological oxygen demand of whey or whey permeate by 75% [43] and aerobic cultivation has reduced BOD by 90–95% [35].

7.2 Environmental considerations and fermentation parameters for *K. marxianus*

Several adjustable factors can influence the rate and quality of fermentation by *K. marxianus*. These factors include the presence of oxygen, nutrient supplementation, substrate, pH and fermentation temperature. In general, hypoxic and anoxic environments favor *K. marxianus's* fermentative metabolism, and ethanol yields are greater than in an aerobic environment [10, 43]. Aerobic conditions favor the building of cell density and are commercially applied for cell propagation in vessels called "Donas" [9]. The doubling time of *K. marxianus* is approximately 70 min, and it has one of the fastest growth rates of any eukaryote [37].

Nutrients are not added to the whey/whey permeate during commercial fermentations [9]. Additional supplementation of nitrogen and phosphorus to whey/whey permeate was shown not to affect ethanol production during fermentation [44]. It has been illustrated experimentally that supplementation of concentrated whey (200 g/l lactose) with bacto-peptone, ergosterol and linoleic acid reduced fermentation time from over 90 to less 60 h [10]. This is a substantial decrease in fermentation time, however large-scale commercial lactose to ethanol fermentations range from 12 to 24 h [8, 9].

K. marxianus is a thermotolerant yeast with reported maximum growth temperatures ranging from 47 to 52°C [10, 35]. Ethanol production has been reported at temperatures as high as 45°C [45] and, other studies indicate that the optimum fermentation temperature is lower. Studies indicate the optimal fermentation temperature range to be 30–40°C [36, 39, 41, 42, 46–48]. This wide range of reported temperatures can likely be attributed to the genetic diversity of *K. marxianus* strains and differences in experimental design. A pH of approximately 5 is widely reported as the optimum fermentation pH value [36, 39, 42, 46–48]. Agitation of fermenting whey/whey permeate occurs in industrial lactose to ethanol conversion and has been incorporated experimentally [8, 9, 47, 49]. To the authors' knowledge, the effects of the rate of agitation on fermentation efficiency of *K. marxianus* have not been examined.

Large scale lactose to ethanol production facilities will adjust fermentation time, temperature, tank pressure, and agitation rate to meet production goals [8, 9]. *K. marxianus* strain UFV-3 may have potential for potable ethanol production. *K. marxianus* strain UFV-3 was able to produce ethanol at yields 90% of the theoretical maximum with fermentation temperatures between 33.3 and 38.5°C, pH 4.7–5.7 and lactose concentrations between 50 and 108 g/l [42].

7.3 Other fermentation organisms

K. lactis is used to produce lactase and recombinant bovine chymosin on an industrial scale. It is the sister organism to *K. marxianus* that is more widely studied. *K. lactis* synthesizes β-galactosidase much like *K. marxianus* and most strains of *K. lactis* are considered Crabtree-negative [50]. A small number of isolate have been used by researchers working with *K. lactis* and it is ubiquitous to fewer environments than *K. marxianus* [10, 32]. This has led to less genetic variation than within in the species than *K. marxianus*. Some strains of *K. lactis* exhibit Crabtree-positive metabolic characteristics [51, 52] and have been genetically engineered for lactose to bioethanol conversion [53]; however, they have not been adopted on a commercial level for lactose to ethanol conversion.

S. cerevisiae is the microorganism widely used in alcoholic beverage and bioethanol production. *S. cerevisiae* is used for traditional potable spirit production for several reasons including its fermentative capacity and ethanol tolerance, being considered Crabtree-positive (preferentially ferments in presence of oxygen and an abundance of glucose), and it's GRAS designation [10]. *S. cerevisiae* is ill-suited for the conversion of lactose to potable ethanol because wild *S. cerevisiae* does not express the genes necessary to produce β-galactosidase. This requires the lactose within whey/whey permeate to be pre-hydrolyzed or *S. cerevisiae* to be genetically engineered to produce β-galactosidase. While pre-hydrolysis of lactose has been explored on an experimental scale for bioethanol production, it would require an additional input (enzymes) and/or additional processing equipment. *S. cerevisiae* preferentially uptakes glucose after lactose hydrolysis and the presence of glucose causes the catabolic repression of enzymes necessary to uptake galactose [54]. The enzymes necessary to uptake galactose will only be synthesized after the glucose has been depleted. This repression causes an increase in fermentation time due to a diauxic lag [10, 55]. While *S. cerevisiae* has been genetically engineered to synthesize β-galactosidase and to reduce catabolic repression, genetically engineered yeast are not commonly used for beverage production [19, 21, 56].

E. coli has been genetically altered to produce ethanol since 1987 [57]. In 2010, *E. coli* was genetically modified to express the *Vitreoscilla* hemoglobin for direct fermentation of sugar to ethanol [58]. This technology has been experimentally developed for the efficient fermentation of whey and other organic by-products. Recently, microbial immobilization has been experimentally applied to *E. coli* expressing *Vitreoscilla* hemoglobin and has shown an increase in lactose to bioethanol production efficiency without producing the microbial biomass associated with the traditional fermentation process [59, 60].

The use of genetically modified organisms for the conversion of whey to potable spirit has the potential to increase production efficiency and reduce operating costs. The use of these organisms will require consumer acceptance of potable spirits produced from this technology.

8. Industrial whey fermentation process and technology for potable spirits

The fermentation process and technology used for the Carbery process are identical for potable spirits and bioethanol production [8, 9]. The Carbery process (**Figure 5**) is used for the industrial conversion of whey to potable spirits. Differences in the process occur during the distillation and during post-distillation processing. Whey/whey permeate is received at the facility and must be cooled to the specified fermentation temperature. Once cooled, the whey is pumped into fermentation tanks and inoculated with *K. marxianus*. The common inoculation rate for commercial spirit production is 1–5 × 10^7 cells/ml [61]. *K. marxianus* is grown in yeast propagation vessels referred to as "Donas". These yeast propagation vessels are aerobic and pumped with filtered air to promote yeast growth. This allows yeast to be maintained in growth phase which increases their ability to produce alcohol and reduces lag time when inoculated in whey [61]. The fermentation tanks are cylindroconical vessels jacketed with ethylene glycol or other coolant for temperature control. The quantity and size of the fermentation tanks vary based upon the production facility capacity. Compressed air is used for agitation during

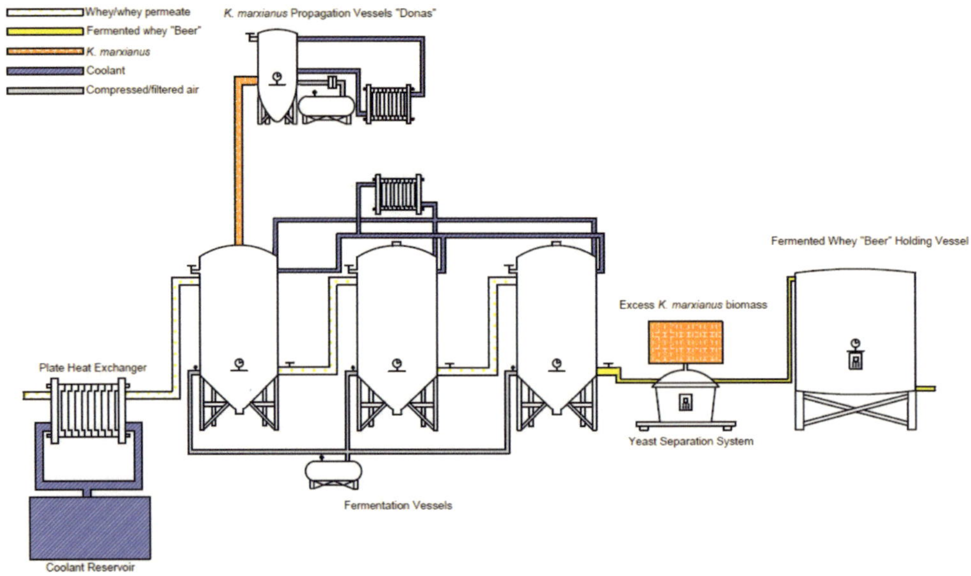

Figure 5.
Example of industrial (Carbery process) whey to potable ethanol fermentation process flow.

fermentation. The fermenting whey is pumped from vessel to vessel with monitoring of the specific gravity occurring throughout the process. The specific gravity of whey starts at approximately 1.022 g/cm^3 and drops to 1.008 g/cm^3 during fermentation. This drop in specific gravity is due the lactose within whey being converted to ethanol and CO_2. The specific gravity measurement is used to determine the process flow rate. Fermentation time ranges between 12 and 24 h [8, 9]. Once the designated specific gravity has been reached, the fermented whey is separated from the yeast via gravity (yeast falling out of solution) and/or through separation technology, such as centrifugation. The yeast can potentially be recycled for later batches or further processed for human or animal consumption. The fermented whey, now called a "beer" is held in a holding tank until distillation [8, 9].

9. Distillation

Once the whey sugars, primarily lactose and its monosaccharide constituents galactose and glucose, have been converted into ethanol, there is a need to concentrate the alcohol up to a strength that is appropriate for a spirituous product. Broadly speaking the ethanol yield from a whey fermentation will be typically 2–5% v/v, depending on the fermentation procedures and any preconcentration applied. The fermented feed though can contain significant levels of other whey constituents such as calcium salts and proteins. Depending on the process design, the whey may be pretreated to remove proteins and salts.

The requirements of the distillation operation are straight-forward, at least in principle. The fermented whey is to a first approximation a dilute solution of ethanol in water, and this ethanol needs to be concentrated by around an order of magnitude to generate the basis of an alcoholic spirit. However, other volatile components present, either from the parent whey or produced during fermentation as secondary metabolites, also termed congeners. Whilst these compounds are present in relatively low concentrations they can contribute to the flavor of the distilled spirit and the distiller needs to make a decision as to how much of these flavors should be retained in the resulting spirit.

In any case, the distillation process consists of three distinct activities: heating, to create vapor from the still feed, condensation, to convert vapor into the liquid spirit, and collection of the spirit. Each of these activities can be achieved using equipment of widely varying complexity and broadly speaking the higher the purity of the alcohol the more complex the equipment needs to be. For the distillation of fermented whey the common primary aim is to create "neutral alcohol" (i.e. alcohol that has no extraneous color or flavor) and so both the concentration of ethanol and removal of flavor-active components is usually required. To achieve this the ratio of surface area to volume in the still is a key design consideration. Generally, the introduction of more surface area tends to enhance the separation of ethanol and congeners, resulting in a cleaner, more neutral spirit. With the rapid development of the craft spirits industry, especially since the turn of the century, there has been a plethora of new still designs and fabricators available to the nascent distiller. To remove any congeners present is usually achieved by multiple distillations, the introduction of "plates" into a still or both.

Whilst the distilled spirit is the primary product from distillation, it is a relatively minor proportion of the still output. If the alcohol is around 3% v/v and the output is, say, 70% v/v, then the spirit fraction is only about 5% of the total feed volume, with the remaining 95% as "waste". However the removal of BOD (mainly present in whey as lactose) and the distillation of ethanol from the fermented whey, means that the BOD is substantially reduced, which in turn reduces effluent costs. If protein is removed prior to distillation and utilized elsewhere, then the resulting still waste stream is amenable to further treatment, for instance by anaerobic digestion. In any case, the distillation operation results in a significant waste stream in itself that must be considered in any process design.

The scale of the fermentation and distillation facilities is straight-forward to estimate. For a cheese plant that produces 5000 kg cheese per year, around 45,000 l of whey will be produced. On a weekly basis this is around 100 kg of cheese and 900 l of whey. If the lactose content is 5% w/v and the sugars are completely fermented (for instance using yeasts such as *K. marxianus*), the ethanol yield will be up to around 3% v/v. Allowing 5 days for a fermentation to complete, two fermenters of 1000 l will be required, and a still of 300–1000 l capacity. The exact capacity depends on how often the distilling operation is performed per week.

10. Still configuration

The recent growth of the craft spirits industry has spawned a wide range of still configurations, many of which focus on flexibility for different feeds. Such stills are referred to as hybrids. As mentioned above, ethanol is only part of the composition of the distilled spirit. A range of other compounds, especially a plethora of esters, short-chain fatty acids and methyl ketones are common secondary metabolites of whey fermentations. Their presence affects the final sensory performance of the spirit and therefore should be under control, either by fermentation management or by judicious distillation.

In principle, most "contaminating" secondary metabolites can be removed by employing four distillation approaches in sequence: stripping, rectification, hydro-extractive distillation and another rectification step. A distiller may not want to remove all the additional flavor-active components. Using a simple pot still, the fermented whey will distil to yield a product of around 15–20% v/v ethanol, depending on the initial ethanol concentration. This ethanol concentration can be increased up to around 70% v/v with a second pot still. This approach will yield a spirit that will retain significant levels of flavor compounds and so will be most "whey-like". If a "cleaner" spirit is required more complexity is required in the distillation set-up.

At the other extreme to the two-pot system is the four-stage system indicated above. Stripping is followed by rectification, a process that typically employs a column of plates to enhance the separation of ethanol from the stripped feed. This should yield an output of close to 96% v/v, close to the maximum concentration of ethanol possible at atmospheric conditions from an aqueous ethanol system (the "azeotropic limit"). But this ostensibly clean spirit still retains flavor from the initial stripping feed and needs further processing to clean up the final spirit. To do this, water is perversely added back to the rectifier column output. This has the effect of increasing the volatility of the secondary metabolites, so that they are more easily separated from the distilling ethanol. The output of this column is still relatively water-rich so an additional rectification stage is the final part of the distillation process to elevate the ethanol concentration toward the azeotropic limit of around 96% v/v.

As mentioned above, there is an option of applying a demethylizer as a final column stage. This is an essential operation for pectin-rich distillation feeds such as those from stone fruits and potatoes. The pectin content of whey is negligible so this is unnecessary. One point to note concerning the use of a demethylizer is that it is most effective at low water concentrations (in contrast to hydro-extractive distillation) and so it is best employed after the second rectification step.

For a plant that only distils whey fermentations, the four-column process has most to commend it, as it will yield spirit that is relatively clean or "neutral". From a craft perspective this is a relatively complex distilling operation (with associated fabrication costs), so novel still configurations are becoming increasingly common. From a customer perspective there are three points to keep in mind when seeking distillation equipment:

- What quality of the final spirit is required in terms of ethanol concentration and levels of secondary metabolites?
- What is the expected range of initial feed ethanol concentration?
- What is the solids content of the original fermented whey feed?

The two former points help to define the distillation stages and the columns that may or may not be required (columns add significant cost to still fabrication). The latter is an important consideration when considering heat source. Direct heating such as electrical elements can be problematic if heating causes precipitation (e.g. of proteins) as they can congeal on to the heating surfaces and can cause heat transfer and burn-on issues for the spirit. The latter in particular can give rise to burnt-on flavors that are difficult to remove from the spirit despite repeated distillations.

11. Use of spirit

A spirit can be used in a range of final distilled spirits. Most commonly, these are vodka, gin and liqueur/cordial products. The specifications for spirit used for vodka production are usually the most exacting. Usually the final product has to be essentially neutral, so that the concentrations of secondary metabolites should be minimal. Typically, spirit for vodka has specifications for total terpenoids, acetic acid, ethyl acetate and methanol. Spirit used for gin must also be neutral, but the use of botanicals to flavor the resulting spirit can help to mask any minor flavor deviations. Liqueurs and cordials based on neutral alcohol are often relatively strong in flavor. In principle a spirit that is less neutralized can be used with relative comfort.

One other aspect to bear in mind is that the addition of sugar, usually as syrups, can help to smooth out any "edges" to the mouthfeel of the spirit. Most liqueurs require significant levels of sugar addition during production, whilst for gin and vodka, only the London dry gin style has proscriptive sugar levels. Returning to the design of the still layout, the decisions there can be steered by the expected uses that the spirit will be put to, with vodka requiring the most tightly defined quality criteria. In any case, though the spirits produced for whatever duty should be of consistent quality.

One other option is to use the spirit for non-potable uses such as fuel. Generally, though the value of a non-potable alcohol product is substantially less than potable alternatives so there is less financial imperative for producing, say, fuel alcohol.

12. Reactive distillation

A relatively recent development in distillation development is the concept of reactive distillation, pioneered by Berglund at Michigan State University. Here the concept is to encourage reactivity between spirit components to alter the sensory attributes of the spirit. This has significant potential value for whey distillates as one demonstrated option is to induce fatty acids to react with ethanol to create esters, mediated by a solid-state acid catalyst. From a whey distillate perspective, this can in principle help to reduce the levels of short-chain fatty acids in spirit (with typical flavor descriptors such as cheesy, rancid) and convert them into fruit-flavored esters. Whilst this has yet to be demonstrated specifically for whey this approach offers a tantalizing option for enhancing the neutrality of whey-derived spirits.

13. Product quality

Spirit quality can be influenced by several factors including source of the whey, fermentation parameters, still configuration and post- distillation product treatment. Congeners, minor volatile constituents of a spirit influence it's the organoleptic qualities. The perception of congeners is considered a flaw in vodka. Congeners are present in raw whey and are formed as secondary metabolites during the fermentation process. Congeners within whey can be carried over during the distillation process and are similar to congeners in other spirits [2].

The source of whey and fermentation parameters can influence the composition of congeners in fermented whey. The composition of the volatile aroma compounds within milk and other dairy products can vary depending upon the source of the milk [62]. The milk producer's diet and geographic location can be attributed to the presence volatile compounds such as terpenes and terpenoids [63, 64]. The cheese production process can also influence volatile compound composition of whey, particularly the application of heat and exposure to microorganisms. Exposure to heat can create thermal artifacts and influence chemical reaction rates within whey. The exposure of milk or whey to microbes can influence the volatile compounds present in whey. The metabolites produced by microbes can include alcohols aldehydes, esters and ketones, all which can influence organoleptic quality. The microbial populations can differ per facility and geographic location [62]. Each cheese production facility can potentially produce wheys with different volatile compound compositions. The source of whey can influence the composition of volatile compounds present in a spirit [11].

Fermentation conditions can influence the production of secondary metabolites of *K. marxianus* [65, 66]. Traditional industrial ethanol production facilities take a *laissez faire* approach to fermentation conditions related to congeners production. These facilities' chief concern is maximizing ethanol production. A similar approach is taken at industrial whey to ethanol production facilities. The extractive distillation process is used to separate ethanol from congeners and produce a neutral spirit. It should be noted that congeners with a similar volatility as ethanol may be more difficult to separate. Diacetyl is difficult to remove via extractive distillation and can impart rancid butter or butterscotch aromas to a spirit [67]. Spirit quality is influenced by the number of plates used to separate the ethanol from the other volatile compounds. A greater number of plates allow for greater separation volatiles reducing the presence of congeners within the final spirit. The use of copper plates or other components which have contact with the spirit during the distillation process can influence the organoleptic qualities of the final product. Copper contact during distillation reduces sulfur aromas in spirits and can reduce concentration of sulfur containing compounds in the final spirit [68]. Post- distillation of filtration of the spirit can reduce the presence of congeners in the final product. Filtration with activated carbon can reduce the congeners in spirits and which can have a perceivable impact on the organoleptic qualities of the spirit [69].

If the spirit is to be sold as a vodka it should have a clean taste with no perceivable aroma. These requirements may not be as stringent if the product is to be sold as flavored spirit or mixed with other ingredients to produce a beverage such as Irish cream. Flavorings may mask presence of congeners or congeners with positive organoleptic qualities may enhance the final product.

For cheese makers with no prior knowledge of distillation, this entire process may appear intimidating. Fortunately, assistance is available for people entering into the distillation business [70].

14. Environmental implications of whey spirit production

The production of potable spirits from whey has the potential to reduce environmental impacts of cheese and spirit production [71]. The fermentation process reduces the environmental impact of whey. The conversion of lactose to CO_2 and ethanol can reduce the BOD of whey by 75% [43] and aerobic cultivation can reduce BOD levels up to 95% [35]. The volume reduction during distillation and reduction of BOD during fermentation indicate that processing spent wash would be less economically and environmentally impactful than raw whey. *K. marxianus* is classified as GRAS and can be used as feed for livestock. It has also illustrated that production of a spirit from whey destined to be land spread instead of a similar grain-based spirit can reduce net CO_2-equivalent emissions [71]. This 2018 study also indicated that the production of a whey-based spirit required less water than a grain-based spirit [71]. These factors indicate that the production of a spirit from whey may be beneficial to the whey producer, distiller and the environment.

15. Conclusion

Whey production can be an economic and environmental problem for small creameries and acid whey producers. The production of potable ethanol from whey is currently occurring on an industrial scale and it may be a strategy worth pursing for smaller producers.

Conflict of interest

The authors do not have any conflict of interest regarding materials covered within this chapter.

Author details

Paul Hughes[1], Derrick Risner[2] and Lisbeth Meunier Goddik[1*]

1 Oregon State University, Corvallis, Oregon, United States of America

2 University of California-Davis, California, United States of America

*Address all correspondence to: lisbeth.goddik@oregonstate.edu

References

[1] Tunick M. Whey processing, functionality and health benefits. In: Onwulata C, Huth P, editors. Whey Processing, Functionality and Health Benefits. 1st ed. Hoboken, NJ: Wiley-Blackwell; 2008. p. 400

[2] Dragone G, Mussatto SI, Oliveira JM, Teixeira JA. Characterisation of volatile compounds in an alcoholic beverage produced by whey fermentation. Food Chemistry. 2009;**112**(4):929-935

[3] Panesar PS, Kennedy JF, Gandhi DN, Bunko K. Bioutilisation of whey for lactic acid production. Food Chemistry. 2007;**105**(1):1-14. Available from: https://www.sciencedirect.com/science/article/pii/S0308814607002816

[4] Dufey PA, Messadene J, Silacci P. Ingestion of whey on alpine pastures by beef cattle and quality of the meat. Agrar Forschung Schweiz. 2016;**7**(1):30-39

[5] Freudenreich P. Carcass value of cows. Effects of feeding sweet whey. Fleischwirtschaft. 1984;**64**(8):958-961

[6] Lynch GP, McDonough FE. USDA research on whey and whey products as feed for cattle. Journal of Agricultural and Food Chemistry. 1979;**27**(4): 695-698. Available from: http://pubs.acs.org/doi/abs/10.1021/jf60224a061

[7] Durham CA, Bouma A, Meunier-Goddik L. A decision-making tool to determine economic feasibility and break-even prices for artisan cheese operations. Journal of Dairy Science. 2015;**98**(12):8319-8332. Available from: https://www.sciencedirect.com/science/article/pii/S0022030215007390

[8] Ling C. Whey to Ethanol: A Biofuel Role for Dairy Cooperatives? Washington, D.C.; 2008. Available from: https://www.rd.usda.gov/files/RR214.pdf

[9] Hamilton R, Wansbrough H. The Manufacture of Ethanol from Whey. Available from: https://nzic.org.nz/app/uploads/2017/10/3H.pdf

[10] Guimarães PMR, Teixeira JA, Domingues L. Fermentation of lactose to bio-ethanol by yeasts as part of integrated solutions for the valorisation of cheese whey. Biotechnology Advances. 2010;**28**(3):375-384. Available from: https://www.sciencedirect.com/science/article/pii/S0734975010000224?via%3Dihub

[11] Risner D, Tomasino E, Hughes P, Meunier-Goddik L. Volatile aroma composition of distillates produced from fermented sweet and acid whey. Journal of Dairy Science [Internet]. 1 Nov 2018;**0**(0). Available from: https://www.journalofdairyscience.org/article/S0022-0302(18)31009-9/fulltext

[12] Blaschek KM, Wendorff WL, Rankin SA. Survey of salty and sweet whey composition from various cheese plants in Wisconsin. Journal of Dairy Science. 2007;**90**(4):2029-2034. Available from: https://www.sciencedirect.com/science/article/pii/S0022030207716910

[13] Outinen M, Heino A, Uusi-Rauva J. Pre-treatment methods of Edam cheese milk. Effect on the whey composition. LWT-Food Science and Technology. 2010;**43**(4):647-654. Available from: https://www.sciencedirect.com/science/article/pii/S0023643809003223

[14] Johansen AG, Vegarud GE, Skeie S. Seasonal and regional variation in the composition of whey from Norwegian Cheddar-type and Dutch-type cheeses. International Dairy Journal. 2002;**12**(7): 621-629. Available from: https://www.sciencedirect.com/science/article/pii/S0958694602000547

[15] Casper JL, Wendorff WL, Thomas DL. Seasonal changes in protein composition of whey from commercial manufacture of caprine and ovine specialty cheeses. Journal of Dairy Science. 1998;**81**(12):3117-3122. Available from: https://www.sciencedirect.com/science/article/pii/S002203029875876X

[16] Belitz H-D, Grosch W, Schieberle P. Carbohydrates. In: Food Chemistry. 4th ed. Berlin, Heidelberg: Springer Berlin Heidelberg; 2009. pp. 248-339. Available from: http://link.springer.com/10.1007/978-3-540-69934-7_5

[17] Belitz H-D, Grosch W, Schieberle P. Milk and dairy products. In: Food Chemistry. 4th ed. Berlin, Heidelberg: Springer Berlin Heidelberg; 2009. pp. 498-545. Available from: http://link.springer.com/10.1007/978-3-540-69934-7_11

[18] Bansal S, Oberoi HS, Dhillon GS, Patil RT. Production of β-galactosidase by *Kluyveromyces marxianus* MTCC 1388 using whey and effect of four different methods of enzyme extraction on β-galactosidase activity. Indian Journal of Microbiology. 2008;**48**(3):337-341. Available from: http://www.ncbi.nlm.nih.gov/pubmed/23100731

[19] Domingues L, Guimarães PMR, Oliveira C. Metabolic engineering of *Saccharomyces cerevisiae* for lactose/whey fermentation. Bioengineered Bugs. 2010;**1**(3):164-171. Available from: http://www.ncbi.nlm.nih.gov/pubmed/21326922

[20] Silva AC, Guimarães PMR, Teixeira JA, Domingues L. Fermentation of deproteinized cheese whey powder solutions to ethanol by engineered *Saccharomyces cerevisiae*: Effect of supplementation with corn steep liquor and repeated-batch operation with biomass recycling by flocculation. Journal of Industrial Microbiology and Biotechnology. 2010;**37**(9):973-982. Available from: http://www.ncbi.nlm.nih.gov/pubmed/20535525

[21] Tahoun MK, El-Nemr TM, Shata OH. A recombinant *Saccharomyces cerevisiae* strain for efficient conversion of lactose in salted and unsalted cheese whey into ethanol. Nahrung/Food. 2002;**46**(5):321-326. Available from: http://www.ncbi.nlm.nih.gov/pubmed/12428446

[22] Schmidt C, Mende S, Jaros D, Rohm H. Fermented milk products: Effects of lactose hydrolysis and fermentation conditions on the rheological properties. Dairy Science and Technology. 2016;**96**(2):199-211. Available from: http://link.springer.com/10.1007/s13594-015-0259-9

[23] Dutra Rosolen M, Gennari A, Volpato G, Volken de Souza CF. Lactose hydrolysis in milk and dairy whey using microbial *β* -galactosidases. Enzyme Research. 2015;**2015**:1-7. Available from: http://www.hindawi.com/journals/er/2015/806240/

[24] Harju M, Kallioinen H, Tossavainen O. Lactose hydrolysis and other conversions in dairy products: Technological aspects. International Dairy Journal. 2012;**22**(2):104-109. Available from: https://www.sciencedirect.com/science/article/pii/S0958694611002251

[25] Das M, Raychaudhuri A. Supply chain of bioethanol production from whey: A review. Procedia Environmental Sciences. 2016;**35**:833-846. Available from: https://www.sciencedirect.com/science/article/pii/S187802961630189X

[26] Roukas T, Lazarides HN. Ethanol production from deproteinized whey by β-galactosidase coimmobilized cells of *Saccharomyces cerevisiae*. Journal of Industrial Microbiology. 1991;7(1):

15-18. Available from: http://link.springer.com/10.1007/BF01575597

[27] Tetra Pak. Whey processing. In: Tetra Pak, editor. Dairy Processing Handbook. Rockford, IL: Tetra Pak; 2018. Available from: http://dairyprocessinghandbook.com/chapter/whey-processing

[28] Coté A, Brown WA, Cameron D, van Walsum GP. Hydrolysis of lactose in whey permeate for subsequent fermentation to ethanol. Journal of Dairy Science. 2004;**87**(6):1608-1620. Available from: https://www.sciencedirect.com/science/article/pii/S0022030204733159

[29] Panesar PS, Kumari S, Panesar R. Potential applications of immobilized β-galactosidase in food processing industries. Enzyme Research. 2010; **2010**:473137. Available from: http://www.ncbi.nlm.nih.gov/pubmed/21234407

[30] Kosseva MR, Panesar PS, Kaur G, Kennedy JF. Use of immobilised biocatalysts in the processing of cheese whey. International Journal of Biological Macromolecules. 2009;**45**(5):437-447. Available from: https://www.sciencedirect.com/science/article/pii/S0141813009002050

[31] Berg JM, Tymoczko JL, Stryer L. Glycolysis and glucogensis. In: Biochemistry. 5th ed. New York: W H Freeman; 2002. pp. 453-496

[32] Fukuhara H. *Kluyveromyces lactis*—A retrospective. FEMS Yeast Research. 2006;**6**(3):323-324. Available from: https://academic.oup.com/femsyr/article-lookup/doi/10.1111/j.1567-1364.2005.00012.x

[33] Lachance M-A. Kluyveromyces van der Walt [1971]. In: The Yeasts. London: Elsevier; 2011. pp. 471-481. Available from: https://www.sciencedirect.com/science/article/pii/B9780444521491000355?via%3Dihub

[34] Ortiz-Merino RA, Varela JA, Coughlan AY, Hoshida H, da Silveira WB, Wilde C, et al. Ploidy variation in *Kluyveromyces marxianus* separates dairy and non-dairy isolates. Frontiers in Genetics. 2018;**9**:94. Available from: http://journal.frontiersin.org/article/10.3389/fgene.2018.00094/full

[35] Fonseca GG, Heinzle E, Wittmann C, Gombert AK. The yeast *Kluyveromyces marxianus* and its biotechnological potential. Applied Microbiology and Biotechnology. 2008;**79**(3):339-354. Available from: http://link.springer.com/10.1007/s00253-008-1458-6

[36] Koushki M, Jafari M, Azizi M. Comparison of ethanol production from cheese whey permeate by two yeast strains. Journal of Food Science and Technology. 2012;**49**(5):614-619. Available from: http://www.ncbi.nlm.nih.gov/pubmed/24082274

[37] Lane MM, Morrissey JP. *Kluyveromyces marxianus*: A yeast emerging from its sister's shadow. Fungal Biology Reviews. 2010;**24**(1–2): 17-26. Available from: https://www.sciencedirect.com/science/article/pii/S1749461310000035

[38] Beniwal A, Saini P, Kokkiligadda A, Vij S. Physiological growth and galactose utilization by dairy yeast *Kluyveromyces marxianus* in mixed sugars and whey during fermentation. 3 Biotech. 2017;**7**(5):349. Available from: http://link.springer.com/10.1007/s13205-017-0985-1

[39] Zoppellari F, Bardi L. Production of bioethanol from effluents of the dairy industry by *Kluyveromyces marxianus*. New Biotechnology. 2013;**30**(6): 607-613. Available from: https://www.sciencedirect.com/science/article/pii/S1871678412008746?via%3Dihub

[40] Grubb CF, Mawson AJ. Effects of elevated solute concentrations on the fermentation of lactose by *Kluyveromyces marxianus* Y-113. Biotechnology Letters. 1993;**15**(6): 621-626. Available from: http://link.springer.com/10.1007/BF00138552

[41] Dragone G, Mussatto SI, Almeida e Silva JB, Teixeira JA. Optimal fermentation conditions for maximizing the ethanol production by *Kluyveromyces fragilis* from cheese whey powder. Biomass and Bioenergy. 2011;**35**(5):1977-1982. Available from: https://www-sciencedirect-com.ezproxy.proxy.library.oregonstate.edu/science/article/pii/S0961953411000602

[42] Diniz RHS, Rodrigues MQRB, Fietto LG, Passos FML, Silveira WB. Optimizing and validating the production of ethanol from cheese whey permeate by *Kluyveromyces marxianus* UFV-3. Biocatalysis and Agricultural Biotechnology. 2014;**3**(2):111-117. Available from: https://www.sciencedirect.com/science/article/pii/S1878818113001035

[43] MIG S. The biotechnological utilization of cheese whey: A review. Bioresource Technology. 1996;**57**(1): 1-11. Available from: https://www.sciencedirect.com/science/article/pii/0960852496000363

[44] Kargi F, Ozmıhcı S. Utilization of cheese whey powder [CWP] for ethanol fermentations: Effects of operating parameters. Enzyme and Microbial Technology. 2006;**38**(5):711-718. Available from: https://www.sciencedirect.com/science/article/pii/S0141022905004916

[45] Kourkoutas Y, Dimitropoulou S, Kanellaki M, Marchant R, Nigam P, Banat I, et al. High-temperature alcoholic fermentation of whey using *Kluyveromyces marxianus* IMB3 yeast immobilized on delignified cellulosic material. Bioresource Technology. 2002;**82**(2):177-181. Available from: https://www.sciencedirect.com/science/article/pii/S0960852401001596?via%3Dihub

[46] Ferreira PG, da Silveira FA, dos Santos RCV, Genier HLA, Diniz RHS, Ribeiro JI, et al. Optimizing ethanol production by thermotolerant *Kluyveromyces marxianus* CCT 7735 in a mixture of sugarcane bagasse and ricotta whey. Food Science and Biotechnology. 2015;**24**(4):1421-1427. Available from: http://link.springer.com/10.1007/s10068-015-0182-0

[47] Ozmihci S, Kargi F. Comparison of yeast strains for batch ethanol fermentation of cheese whey powder [CWP] solution. Letters in Applied Microbiology. 2007;**44**(6):602-606. Available from: http://doi.wiley.com/10.1111/j.1472-765X.2007.02132.x

[48] Hadiyanto AD, Aini AP, Pinundi DS. Optimization of ethanol production from whey through fed-batch fermentation using *Kluyveromyces marxianus*. Energy Procedia. 2014;**47**: 108-112. Available from: https://www.sciencedirect.com/science/article/pii/S1876610214002197

[49] Ozmihci S, Kargi F. Kinetics of batch ethanol fermentation of cheese-whey powder [CWP] solution as function of substrate and yeast concentrations. Bioresource Technology. 2007;**98**(16):2978-2984. Available from: https://www-sciencedirect-com.ezproxy.proxy.library.oregonstate.edu/science/article/pii/S0960852406005232

[50] Rodicio R, Heinisch JJ. Yeast on the milky way: Genetics, physiology and biotechnology of *Kluyveromyces lactis*. Yeast. 2013;**30**(5):165-177. Available from: http://www.ncbi.nlm.nih.gov/pubmed/23576126

[51] González Siso MI, Ramil E, Cerdán ME, Freire-Picos MA. Respirofermentative metabolism in *Kluyveromyces lactis*: Ethanol production and the Crabtree effect. Enzyme and Microbial Technology. 1996;**18**(8): 585-591. Available from: https://www.sciencedirect.com/science/article/pii/0141022995001514

[52] El-Batal AI, Farahat LM, El-Rehim HA. Ethanol production by *Kluyveromyces lactis* immobilized cells in copolymer carriers produced by radiation polymerization. Acta Microbiologica Polonica. 2000;**49**(2): 157-166. Available from: http://www.ncbi.nlm.nih.gov/pubmed/11093678

[53] González-Siso MI, Touriño A, Vizoso Á, Pereira-Rodríguez Á, Rodríguez-Belmonte E, Becerra M, et al. Improved bioethanol production in an engineered *Kluyveromyces lactis* strain shifted from respiratory to fermentative metabolism by deletion of NDI1. Microbial Biotechnology. 2015;**8**(2): 319-330. Available from: http://www.ncbi.nlm.nih.gov/pubmed/25186243

[54] Gancedo JM. Yeast carbon catabolite repression. Microbiology and Molecular Biology Reviews. 1998;**62**(2):334-361. Available from: http://www.ncbi.nlm.nih.gov/pubmed/9618445

[55] Mehaia MA, Cheryan M. Ethanol from hydrolyzed whey permeate using *Saccharomyces cerevisiae* in a membrane recycle bioreactor. Bioprocess Engineering. 1990;**5**(2):57-61. Available from: http://link.springer.com/10.1007/BF00589146

[56] Ostergaard S, Olsson L, Nielsen J. Metabolic engineering of *Saccharomyces cerevisiae*. Microbiology and Molecular Biology Reviews. 2000;**64**(1):34-50. Available from: http://www.ncbi.nlm.nih.gov/pubmed/10704473

[57] Kambam PKR, Henson MA. Engineering bacterial processes for cellulosic ethanol production. Biofuels. 2010;**1**(5):729-743. Available from: https://www.tandfonline.com/doi/full/10.4155/bfs.10.46

[58] Sanny T, Arnaldos M, Kunkel SA, Pagilla KR, Stark BC. Engineering of ethanolic *E. coli* with the *Vitreoscilla* hemoglobin gene enhances ethanol production from both glucose and xylose. Applied Microbiology and Biotechnology. 2010;**88**(5):1103-1112. Available from: http://link.springer.com/10.1007/s00253-010-2817-7

[59] Sar T, Seker G, Erman AG, Stark BC, Yesilcimen Akbas M. Repeated batch fermentation of immobilized *E. coli* expressing *Vitreoscilla* hemoglobin for long-term use. Bioengineered. 2017; **8**(5):651-660. Available from: https://www.tandfonline.com/doi/full/10.1080/21655979.2017.1303024

[60] Sar T, Stark BC, Yesilcimen Akbas M. Effective ethanol production from whey powder through immobilized *E. coli* expressing *Vitreoscilla* hemoglobin. Bioengineered. 2017;**8**(2):171-181. Available from: https://www.tandfonline.com/doi/full/10.1080/21655979.2016.1218581

[61] Kelsall DR, Lyons TP. Practical management of yeast: Conversion of sugars to ethanol. In: Jacques KA, Lyons TP, Kelstall DR, editors. Alcohol Textbook. 4th ed. Nottingham: Nottingham University Press; 2003. pp. 121-133

[62] Turbes GS. Impact of Terroir on Cheddar Cheese Flavor and the Influence of Farm-to-Farm Variability, Commingling, and Pasteurization. Oregon State University; 2014. Available from: https://ir.library.oregonstate.edu/concern/graduate_thesis_or_dissertations/7p88ck174?locale=en

[63] Lejonklev J, Løkke MM, Larsen MK, Mortensen G, Petersen MA, Weisbjerg

MR. Transfer of terpenes from essential oils into cow milk. Journal of Dairy Science. Champaign, IL; 2013;**96**(7): 4235-4241. Available from: https://www-sciencedirect-com.ezproxy.proxy.library.oregonstate.edu/science/article/pii/S0022030213003123

[64] Borge GIA, Sandberg E, Øyaas J, Abrahamsen RK. Variation of terpenes in milk and cultured cream from Norwegian alpine rangeland-fed and in-door fed cows. Food Chemistry. 2016; **199**:195-202. Available from: https://www-sciencedirect-com.ezproxy.proxy.library.oregonstate.edu/science/article/pii/S0308814615302430

[65] İşleten Hoşoğlu M. Study of increasing the production of volatile flavor compounds by the yeast *Kluyveromyces marxianus* through optimization of carbon and nitrogen sources. Food and Health. 2018;**4**(2): 112-123. Available from: www.scientificwebjournals.com

[66] Wittmann C, Hans M, Bluemke W. Metabolic physiology of aroma-producing *Kluyveromyces marxianus*. Yeast. 2002;**19**(15):1351-1363. Available from: http://doi.wiley.com/10.1002/yea.920

[67] Walker G, Hill A. *Saccharomyces cerevisiae* in the production of whisk[e]y. Beverages. 2016;**2**(38):1-15. Available from: file:///C:/Users/derri/Downloads/beverages-02-00038[3].pdf

[68] Harrison B, Fagnen O, Jack F, Brosnan J. The impact of copper in different parts of malt whisky pot stills on new make spirit composition and aroma. Journal of the Institute of Brewing. 2011;**117**(1):106-112. Available from: http://doi.wiley.com/10.1002/j.2050-0416.2011.tb00450.x

[69] Siříšťová L, Šárka P, Riddellová K, Hajšlová J, Melzoch K. Changes in quality parameters of vodka filtered through activated charcoal. Czech Journal of Food Sciences. 2012;**30**:474-482. Available from: https://www-cabdirect-org.ezproxy.proxy.library.oregonstate.edu/cabdirect/FullTextPDF/2012/20123285582.pdf

[70] Risner D, Hughes P, Meunier-Goddik L, Ramirez S. Starting a Distilled Spirit Business: A Guide to Completing the Alcohol and Tobacco Tax and Trade Bureau Application and the Oregon Liquor Control Commission Licensing Process [Internet]. 1st ed. Corvallis: Oregon State University Extension Service; 2018. p. 29. Available from: http://sos.oregon.gov/business/Pages/default.aspx

[71] Risner D, Shayevitz A, Haapala K, Meunier-Goddik L, Hughes P. Fermentation and distillation of cheese whey: Carbon dioxide-equivalent emissions and water use in the production of whey spirits and white whiskey. Journal of Dairy Science. 2018; **101**(4):2963-2973

Chapter 3

Glycomacropeptide: Biological Activities and Uses

Juan Jauregui-Rincón, Eva Salinas-Miralles, Norma Chávez-Vela and Mariela Jiménez-Vargas

Abstract

Glycomacropeptide (GMP) is a milk-derived bioactive peptide that comprises 15–20% of proteins present in whey, being the third most abundant. It is released from κ-casein by enzymatic digestion, either physiologically or in industry during cheese making process. GMP has many biological activities that are of particular interest for the manufacture of novel functional foods. Specifically, health promoting activities of this whey peptide are related to: antimicrobial, anticariogenic, gastric acid inhibitory, cholecystokinin releasing, prebiotic, and immune modulatory. GMP is also a peptide with promising use in food industry, due to its nutritional value and its emulsifying, foaming, and gelling properties. Besides, GMP has received much attention due to its use as an indicator of milk adulteration with cheese whey. This chapter summarizes the current knowledge about biological activities of GMP, going in-depth in immune regulatory properties, exposes the potential uses of GMP in industry, and finally reviews different methods used to detect GMP as adulteration index with cheese whey.

Keywords: glycomacropeptide, bioactive peptide, health promoting activities, whey component, milk adulteration

1. Introduction

As already mentioned in other chapters, milk whey is a liquid by-product generated after obtaining cottage cheese or curd (proteins coagulated by acid and heat), also known as cheese whey, that for many years has been considered a waste product, and sent to bodies of water, soil, and sewage systems. However, currently it is used due to its multiple nutritional and functional properties [1].

In Mexico, the production of whey in 2016 was estimated at 1,010,000 tons, 47% of which was discharged to soil, drains, and bodies of water. Despite the fact that multiple uses have been found to cheese whey, this has become a serious environmental problem [2]. This by-product is composed of water, lactose, proteins, peptides, fat, and mineral salts [3]. One of the peptides of interest is glycomacropeptide (GMP), which is obtained after the coagulation of milk κ-casein during cheese production and represents 15–20% (w/w) of the total proteins contained in milk whey [4].

GMP is the C-terminal fragment released by the proteolytic action of the endopeptidase chymosin (renin) on κ-casein during the initial stages of cheese making, or by the action of pepsin during the gastric digestion. κ-casein is hydrolyzed at phenylalanine105-methionine106 bond, forming two very different polypeptides. One is called para-κ-casein (residues 1–105), and it is slightly cationic at pH 6.6, hydrophobic and poorly soluble, which remains in cheese curd; and the other is GMP (residues 106–169), that is strongly polar so diffuses into the aqueous phase, being eliminated during the draining with the cheese whey (as reviewed in [5]).

2. Chemical properties and molecular structure of GMP

GMP has 64 amino acid residues, with an isoelectric point (pI) between 4 and 5. Fifty percent of GMP is deglycosylated and is known as caseinomacropeptide (CMP) [5]. However, milk GMP can present different types of carbohydrates, such as: sialic acid, galactosyl, and N-acetylgalactosamine, which generate different glycosylated forms of the molecule. GMP is rich in amino acids such as proline, glutamine, serine, and threonine, but deficient in tryptophan, tyrosine, phenylalanine, and cysteine. The absence of aromatic amino acids in its primary structure causes that GMP does not present absorption at the wavelength of 280 nm. However, GMP can be detected at wavelengths between 205 and 226 nm and absorption differences between 210 and 280 nm are used for the characterization of GMP (as reviewed in [5]). The composition of GMP can be variable and depends on the source of serum and the fractionation technology used in its isolation [3] (**Figure 1**).

As reviewed by Neelima [7], the three-dimensional structure of GMP cannot be evaluated due to its crystallization which is not possible, so it can only be seen from a purely theoretical approach. GMP is a peptide that does not possess defined secondary and tertiary structure. However, three-dimensional structure of GMP has been predicted by means of protein modeling and shows that a large part of the peptide has a strong negative charge, whereas there are three small domains with a positive charge at the N-terminal end. At pH 7.0, its mean value of the hydropathy is −0.322, and GMP is more hydrophilic than hydrophobic. The hydropathy value decreases when glycosylation of GMP increases, due to the greater amount of sialic acid residues.

101 H . Met—Ala—Ile—Pro—Pro—Lys—Lys—Asn—Gln—Asp—Lys—Thr—Glu—Ile—Pro

Ile (Variant B)

121 Thr—Ile—Asn—Thr—Ile—Ala—Ser—Gly—Glu—Pro—Thr—Ser—Thr—Pro—Thr—Thr—Glu—Ala—Val—Glu

Ala (Variant B)

141 Ser—Thr—Val—Ala—Thr—Leu—Glu—Thr—Ser—Pro—Glu—Val—Ile—Glu—Ser—Pro—Pro—Glu—Ile—Asn

161 Thr—Val—Gln—Val—Thr—Ser—Thr—Ala—Val. OH

Figure 1.
Primary structure of bovine GMP variant A and B, where ● indicates its three phosphorylation sites and ▲ the most important glycosylation sites. Modified from Thomä-Worringer et al. [6].

3. Biological activities of GMP

The use of GMP is growing, since it is a bioactive peptide with unique nutritional and nutraceutical properties. Many biological activities of GMP have been reported, highlighting antimicrobial, anticariogenic, gastric acid inhibitory, cholecystokinin (CCK) releasing, prebiotic, and immune modulatory. Of particular interest is GMP's capacity to modulate the immune response, due to its potential use in treatment or prevention of different immunopathologies.

One of the first antimicrobial effects observed in GMP was due to its ability to bind cholera toxin and *Escherichia coli* enterotoxins. Chinese hamster ovary (CHO)-K1 cells undergo morphological changes in presence of cholera toxin and GMP at 20, 100, and 1000 μg/mL was able to suppress this morphological change by more than 70%. Treatment of GMP with proteinases lowered this activity, but removal of sialic acid abolished it [8]. Curiously, as GMP doses increased the inhibitory effect decreased. Authors demonstrated that sialic acid is mediating this inhibitory activity, so it could be inferred that GMP at high doses has less sialic acid available probably as consequence of its aggregation into polymers. Likewise, GMP has showed inhibitory activity on CHO-K1 cells morphological change induced by *E. coli* thermolabile enterotoxins [9]. Later, the binding ability of GMP to intestinal pathogenic bacteria was evidenced, mainly to entherohemorragic *E. coli* (EHEC O157). This activity decreased when the peptide was desialylated and peroxidated, showing that sialic acid is essential to GMP attachment. Besides, GMP prevented in a dose-dependent manner the adhesion of EHEC O157 to Caco-2 cells and when it was conjugated with xylooligosaccharide or carboxymethyldextran, the release of IL-8 by Caco-2 was also suppressed [10]. The same inhibitory effect on pathogenic bacteria adhesion to Caco-2 cells was corroborated with 3[H] thymidine-labeled enterophatogenic *E. coli* (EPEC) *Salmonella typhimurium* or *Shigella flexneri* [11]. Besides, it has been reported that GMP generates dose-dependent inhibition of enterotoxigenic *E. coli* (ETEC) K88 adhesion to ileal mucosa *in vitro* [12]. Finally, a recent report showed that GMP also prevents the attachment of several strains of EHEC and EPEC to Caco-2 and HT-29 mammalian cells [13]. In this study, it was demonstrated that GMP, beyond inhibiting bacterial adhesion, is able to maintain the structural integrity of tight junctions on Caco-2 monolayers, thus delaying the paracellular translocations of EPEC. However, it is important to clarify that authors did not measure the expression level of proteins associated with these junctions.

There are several *in vivo* assays that show a probable protective effect of GMP against pathogenic bacteria. GMP has been reported to neutralize in 80% the rate of incidence of diarrhea induced by cholera enterotoxins LT-1 in mice, and abolished that induced by LT-2 [9]. On the other hand, the addition of GMP into diets of piglets diminished the bacterial count in mucosal scrapping and abated the attachment of ETEC K88 to the intestine, mainly on the ileal mucosa where the receptor for *E. coli* is located [12]. Moreover, a diet supplemented with 1% GMP protected weaning piglets from damage caused by *E. coli* infection, and prevented the reduction of growth, morphological damage, and the increase in intestinal permeability associated to infection [14].

In association with this antimicrobial effect, an anticariogenic activity to GMP has been demonstrated. Firstly, *in vitro* assays using different products that contain GMP as an active principle showed that it could inhibit the adhesion of bacteria inducing dental plaque and carious in oral cavity to surface plastic, such as *Streptococcus mutans*, *S. sanguis*, and *Actinomyces viscosus* [15]. Later assays corroborated these results, as the incorporation of GMP with or without caseinophosphopeptide to salivary films modified the adhesion capacity of *S. sobrinus* and *S. mutans* to bovine enamel discs [16]. This antimicrobial effect was linked to a remineralization activity of GMP, as demonstrated later through an experimental protocol in human that showed that GMP alone or combined with xylitol promotes more remineralization than commercial fluoride toothpaste [17]. The anticariogenic and remineralization activity of GMP have been claimed in several patents [15, 17].

Several studies have related GMP with the inhibition of gastric secretion. First ones were mostly developed using dogs by a group of Russian researchers. The first evidence that GMP inhibits gastric secretion was showed by Shlygin and co-workers [18] using gastrin to evoke it. Subsequent works demonstrated similar effect using different gastric secretion stimulants [19]. Some years later, it was proposed that this inhibitory effect was caused by a GMP fragment rather than the whole molecule [20, 21]. Later, injecting dogs with a protein fraction obtained from the gastric content of unweaned rats, it was observed an inhibition in dog gastric secretion to a food stimulus [22]. This inhibitory action was similar to that induced by GMP in dogs. GMP was also demonstrated to inhibit gastric motility after its intravenous injection in dogs [23]. All these experiments point out that at physiological conditions GMP may be playing a crucial role in the preservation of active milk proteins in newborn animal during natural breast feeding. In addition to dogs, other experimental models such as rats, pigs, and calves and also isolated organs were used to demonstrate that GMP induces gastric secretion inhibition in association with a decrease in blood of some regulatory digestive hormones, as gastrin and CCK (as reviewed in [24]). However, variations in used gastric stimuli, GMP dose, and origin, via of administration and experimental approach may be the cause of the differences in the reported intensity to this GMP activity.

Related with the effect of this bioactive peptide on digestive hormones, GMP has also been associated with appetite control. Several *in situ* studies with Wistar rats have suggested that glycosylated forms of GMP A variant could regulate food intake through CCK secretion, a hormone involved in satiety, as GMP stimulate its release [25, 26]. Nevertheless, in studies in human, GMP had no effect in the regulation of food intake over a short-term period [27], neither in the loss of body weight after 12 months of sustained consumption [28]. Likewise, GMP with different degree of glycosylation does not modify the concentration of CCK in human plasma [29]. As other authors have pointed out [27], these inconsistencies related to GMP's effect on CCK release are probably due to dose changes in both animal models and human trials.

For many years, the prebiotic properties of GMP have been discussed. The first evidence that GMP might possess prebiotic activity arose with the bifidobacterial growth promoting effect of human's colostrums and milk by *in vitro* assays [30]. In an attempt to discern the component of the milk responsible for this activity, an important role of N-acetyl-glucosamine (GlcNAc) and oligosaccharides with GlcNAc [31] was pointed out. Subsequent investigations were quite contradictory, until the prebiotic activity of GMP was demonstrated for the first time in 1984 [32]. This work showed that human GMP is a promoter of bifidobacterial growth, although the effect was lost when it was hydrolyzed, as well as when bovine GMP was used, showing the importance of the GMP peptide chain to function as a prebiotic. Eight years later, it was reasserted that peptide fraction was decisive for the prebiotic effects on bifidus [33]; while other researchers leaned by sialic acid as the inductor of the effect [34]. Following this line, the supplementation of milk with 2% GMP increased the *in vitro* growth of *Bifidobacterium lactis* [35]. Besides, the addition of 2 mg/mL of cow GMP to the growth medium, promoted the growth of *Lactobacillus rhamnosus* and *Bifidobacterium thermophilum*, and apparently glycosylation was not an essential factor to carry out this function [36]. On the other hand, using an artificial colon model to simulate colonic fermentation, GMP was shown to modulate the gut microbiota of elderly subjects by promoting the growth of several health-relevant taxa, like *Coprococcus* and *Dorea*, both related to resistance to pathobiont colonization [37]. Finally, a recent *in vitro* research has shown that GMP promotes the growth of *Bifidobacterium longum* ssp. *infantis* in a dose-dependent manner and modulates its genes expression. Again, they found that the effect was lost with periodate GMP, suggesting that its activity is due to the oligosaccharides present in the molecule [38].

In the last years, several research groups have demonstrated that oral treatment with GMP modifies *in vivo* the microbiota in gut using different experimental approaches. First works were developed in mice, and showed that after 15 days of

GMP treatment, there was a significant increase in *Lactobacillus* and *Bifidobacteria* in fecal samples, at the same time the number of *Enterobacteriaceae* and coliforms decreased [39]. Feeding mice for 8 weeks with a GMP-enriched diet reduced *Desulfovibrio* bacteria in normal and phenylketonuric mice; a bacteria that is associated with the pathogenesis of inflammatory bowel disease (IBD). Likewise, normal mice increased Firmicutes and phenylketonuric ones Bacteroidetes; specifically, the genus *Allobaculum* (associated with body weight loss [40]) and *Bacteroidales*, respectively [41]. In this case, the prebiotic property of GMP was associated with an increase in short chain fatty acid production, mainly acetate and propionate, that may has an important role in the regulation of immune response. So, although the prebiotic activity of GMP has been demonstrated, more research is needed to clarify whether the peptidic or carbohydrate fraction or both are involved in this bioactivity.

3.1 The immunomodulatory properties of GMP

GMP has been shown to modulate the immune response in a number of different ways. First, we summarize literature reports about regulatory activity of GMP on immune cells demonstrated by *in vitro* assays, and later, we will focus on those studies in which the regulatory effect of GMP on immune response was analyzed in animal models.

In relation to the immunomodulatory effects of GMP on immune cells, different *in vitro* approaches have corroborated the inhibitory action of GMP on splenocyte proliferative response to mitogens, and have shown that both sialic acid residues and polypeptide portions of GMP are essential in this inhibitory effect. In 1992, it was showed that the GMP fraction obtained from κ-casein inhibited proliferation of mouse splenocytes induced by *Salmonella typhimurium* lipopolysaccharide (LPS) [42]. Furthermore, GMP displayed an inhibitory activity on the proliferative response induced by concanavalin A (Con A) and phytohemagglutinin (PHA) [42, 43]. Initially, it was found that sialic acid was the key in this inhibitory activity, as it was lost after digestion with neuraminidase [44]. However, the inhibitory effect of GMP was increased after digestion with trypsin and pronase, which suggests that the peptidic chain is also involved in this immunomodulatory activity. In this regard, the same working group showed that the inhibitory effect on PHA-induced proliferative response is higher when the number of sialic acid residues is increased and that on LPS-induced proliferation is highest with a GMP fraction containing two sialic acid residues [45]. Both inhibitory effects decreased significantly after neuraminidase digestion. They also suggested that phosphate group at serine-149 plays a role in GMP binding to the mitogen receptor, as they observed a reduced inhibitory activity after GMP chymotrypsin digestion. Regarding to the associated mechanism to GMP inhibitory effect on splenocyte proliferation, it was showed that GMP stimulates the synthesis of a soluble inhibitory component, an interleukin (IL-1) receptor antagonist or IL-1ra [46, 47]. Moreover, GMP was able to bind to mouse CD4+ helper T cells and to suppress the expression of the IL-2 receptor on the cell membrane, inhibiting the PHA-induced proliferation of mouse splenocytes [48]. Subsequently, the inhibitory action of GMP on LPS-induced cellular proliferation was confirmed in mouse splenocytes, although they did not report any effect on PHA- or Con A-stimulated cells [49]. But later, controversy about the effect of GMP on the *in vitro* proliferation of spleen cells was generated, as it was reported that GMP increases the proliferation response of lymphocytes stimulated by Con A [50]. In this study, an increase in Foxp-3 and IL-10 expression was also demonstrated. Besides, authors showed an inhibition on secretion of IFN-γ and TNF-α and on STAT4 activation when cells were stimulated by Con A in presence of GMP. The same team studied the action of GMP on monocyte cell line THP-1 and they found that GMP increases the secretion of TNF, IL-1β, and IL-8 by THP-1 cells, and this effect is mediated via MAP kinase and NF-kB pathways [51].

On the other hand, GMP is also able to downregulate dendritic cell response to LPS by inducing a slight but significant decrease in the production of IL-6, IL-1β, and TNF-α, but without changing the production of IL-12 and IL-10 [49]. Strikingly, the regulatory effect of GMP on neutrophils is the opposite, as it improves proliferation and phagocytic activity of the human macrophage like cells U937 [52]. However, the observation that both polypeptide and carbohydrate portions are essential for GMP biological effects is reinforced in this study, as peptides of pepsin-digested GMP and sialic acid-rich GMP fractions significantly enhanced cell proliferation and phagocytic activities stimulated by non-digested or asialo-GMP on U937 cell. Also, an upregulatory effect of GMP on production of IgA by LPS-stimulated splenocytes has been reported, being correlated with an increase in the population of IgA positive cells [53].

There are several studies that analyze the immunomodulatory activity of GMP on immune response when it is orally administered to experimental animals. In the context of splenocytes response to mitogens, two *in vivo* studies were carried out to analyze the possible immunomodulatory activity of GMP. First one was developed in 1998 and demonstrated that mice fed with a GMP-supplemented diet show an enhanced proliferative response of spleen cells to Con A, without generating significant changes in the response to LPS or PHA [54]. Later in 2012, it was showed that oral intake of GMP by rats reduces the proliferative response of splenocytes induced by Con A [55]. In both *in vivo* studies, animals were antigen-immunized because antibody response was also measured. All together *in vitro* plus *in vivo* studies, point out the inhibitory effect of GMP on splenocyte proliferation to mitogens. The opposite response reported by one *in vitro* [50] and *in vivo* [54] assay was quite possibly due to concentration-dependent effects or assay-used conditions.

The effect of orally administered GMP on humoral immunity has also been studied. Mice fed with GMP have shown suppressed levels of specific IgG to dietary and injected antigens, with no change in IgM, IgA, and IgE antibody response [54]. In this regard, a recent study showed that oral administration of GMP to mice resulted in a greater number of IgA positive plasma cells in the intestinal lamina propria [56]. All these results [54, 56] plus *in vitro* ones [53] about Igs production fit together, suggesting an immuno-suppressing activity of GMP on systemic humoral response, but an immuno-stimulating activity on humoral mucosal immunity.

Martínez-Augustin and co-workers [57, 58] have studied the immunomodulatory action of GMP in experimental models of intestinal inflammation. They have demonstrated that GMP administered orally to rats exerts an anti-inflammatory effect in ileitis and colitis induced with trinitrobenzenesulfonic acid (TNBS); said anti-inflammatory effect shows a degree of efficacy similar to that of sulfasalazine, a drug widely used in the treatment of inflammatory bowel disease. GMP was shown to protect rats from TNBS-induced colonic and ileal inflammatory damage, by reducing the damage score and the extent of necrosis, and also by diminishing the increased alkaline phosphatase colonic activity and inducible oxide nitric synthase expression. IL-1β and IL-1ra messenger RNA levels were significantly decreased in colon as a consequence of GMP administration; and myeloperoxidase activity and levels of IL-1β and IL-17 were decreased in ileum. Initially, the authors assumed that the action mechanism of GMP was not related to anti-oxidative activity or to regulatory cell induction, as glutathione or TGF-β levels in colon and Foxp-3 in ileum were not affected [57, 58]. However, when GMP was orally administered to rats, an increase on Foxp3 expression in spleen cells was observed, although secretion of cytokines by *ex vivo* Con A-stimulated splenocytes did not change [50]. Putting together these results with the regulatory activity of

GMP on monocytes (THP-1) and splenocytes cytokine response obtained by the same working group and previously mentioned in this review [50, 51], authors concluded that the intestinal anti-inflammatory action of GMP is likely to be mediated by the direct modulation of monocyte or splenocyte activity, especially by hampering the activation of Th1 cells while favoring the differentiation of Treg cells [50].

In recent years, a Mexican laboratory led by Salinas [55, 59–61] has focused on the study of the immunomodulatory activity of GMP in experimental allergy models. They found that oral administration of GMP to rats before and during sensitization with allergen significantly reduces the level of allergen-specific IgE in serum, and also decreases the proliferative response and the production of IL-13 by splenocytes stimulated by the allergen [55]. Treatment of animals with GMP also protected them from systemic anaphylaxis as GMP administration increased survival rates and lessened signs of severity of anaphylactic shock. Moreover, GMP reduced the intensity of urticarial inflammatory reaction when sensitized animals were intradermically challenged with the allergen [55]. With these results, it was demonstrated the immunomodulatory properties of GMP on allergic sensitization and its beneficial effect on clinical signs associated to early-phase allergic reaction. Then, they investigated whether GMP may impact on late-phase and chronic inflammatory allergic reactions, using two experimental models that after repetitive exposure to allergens displayed local recruitment and activation of immune cells with persistent production of inflammatory mediators in affected tissues, together with substantial changes in the extracellular matrix and alterations in structural cells [62]. Specifically, they used experimental models of asthma and atopic dermatitis prophylactically administered with GMP, that is to say, prior to and during pathology establishment. As expected, GMP intake resulted in reduction of IgE titers in serum. In addition to this, in asthma model, GMP substantially decreased blood eosinophilia and suppressed the recruitment of inflammatory cells to the bronchoalveolar compartment. GMP also inhibited eosinophils infiltration, goblet cells hyperplasia, and collagen deposit in lung tissue [59]. Equivalent results were obtained in allergen-induced atopic dermatitis model, where GMP reduced the intensity of cutaneous inflammatory process and edema, abolished pruritus, and reduced eosinophils recruitment and mast cells hyperplasia in dermis [60]. In both models, expression of IL-5 and IL-13 was markedly inhibited in lung and skin, while expression of IL-10 was increased. Their research then turned to the mechanism by which GMP modulates the allergic response. They demonstrated that GMP administration increases the amount of *Lactobacillus* and *Bifidobacterium* present in gut of allergen-sensitized animals after 3 days of oral treatment, and that of *Bacteroides* after 17 days. Interestingly, this intestinal microbiota is associated with protection in allergy. GMP intake also increased the production of TGF-β by splenocytes of sensitized animals in response to allergen and impacted mast cell function, inhibiting their activation and also the release of histamine in response to allergens. No change in tissue mast cell number was found [61]. These results obtained in experimental allergy models again show a double way by means GMP exerts its control on inflammation; on one hand, through a direct modulation of immune cells activity involved in the process, and on other side by potentiating a regulatory microenvironment against the Th2-inflammatory one. More studies are needed to understand which immune cell receptors recognize GMP and which intracellular signals activate or inhibit.

Finally, there are few studies that analyze the role of GMP on cancer. In a rat model of pharmacological-induced colorectal cancer, oral administration of 100 mg/kg of GMP decreased the number of aberrant crypt foci although no effect was observed at doses of 10 and 50 mg/kg. On the other hand, there was no change in methylation and expression level of p16 and MUC2, two tumor suppressor genes [63]. Additionally, through an *in vitro* assay GMP was showed to inhibit the expression of p65 NF-kB in human colorectal tumor HT-29 cells activated with LPS, key element in colorectal cancer induced by inflammatory bowel diseases [64].

Although more studies are needed in relation to some biological activities, current results propose GMP as a good candidate to be used as a functional ingredient in food industry.

4. Potential uses of GMP

Today, one of the objectives of the food industry is the development of novel food products with beneficial properties for health. For its different health benefits, GMP can be used in therapeutic and dietary foods, or as a functional ingredient in various special products, like oral care products.

4.1 Therapeutic and nutritional applications

It is crucial to demonstrate that GMP is hypoallergenic to be used in food compositions. In this regard, Takahashi and collaborators patented a food composition that contained GMP and a mixture of free amino acids (leucine, lysine, methionine, cysteine, phenylalanine, tyrosine, tryptophan, arginine, histidine, and glycine) [65]. The composition presented good taste, good absorption and digestion properties, and a high nutritional value. They demonstrated that this composition was hypoallergenic, as after repeated injections of the GMP composition together with an adjuvant used to induce experimental allergy in mice, no antibody against GMP was detected in serum by Ouchterlony. Although this method is not very accurate, GMP hypoallergenicity was later corroborated by Milkkelsen and collaborators using ELISA test to show absence of specific antibodies in mice after being sensitized both systemically or orally with GMP [66].

Due to the particular amino acid composition of GMP, devoid of aromatics amino acids (phenylalanine, tryptophan, and tyrosine), it can be used for special diets of people suffering from phenylketonuria (PKU), being an adequate choice as a source of proteins [67]. On the other hand, GMP has low amount of methionine but high amount of branched chain amino acids (valine and isoleucine), which makes this peptide an excellent candidate to be used for the control of liver diseases, as this type of amino acids are good as caloric sources [68]. There is a patent to use of GMP to improve female's health [69]. The inventors claim that administration of a composition comprising GMP can improve the health of the females. They used murine models fed with GMP composition and showed that females decreased final fat mass and percent body fat, when comparing with females that received a diet based on caseins or free amino acids as source of proteins. In relation to bone characteristics, femur length was larger in GMP administered mice, although only females showed less femoral weakness and greater bone mineral content and density as compared to those fed with amino acids or casein diets, respectively.

4.2 Dietary supplementation

As previously mentioned, research results suggest that GMP has an effect on the feeling of fullness but this does not translate into a lower food intake [27, 28]. For an application in food intake regulation and in potentially body weight management, more work is required. Understanding dose, timing, and delivery mode, including food form and composition, in relation to the pattern of release of CCK, is needed for the use of GMP as appetite suppressant [70].

4.3 Food additive

GMP has physicochemical properties that make it attractive for use as an additive in food products. According to studies on the functional properties of GMP, it can act as an emulsifier, foaming, and gelling agent.

GMP as an emulsifier presents stability to pH variations, which is attractive for foods that undergo pH changes during their process, such as the case of fermented milk products [71]. The best emulsifying capacity was obtained at alkaline pH. However, it has been observed that emulsions with GMP as emulsifying agent are not stable during storage when they have received thermal treatment [72]. Besides, GMP modified covalently with disaccharides or fatty acids can present an improved function and even increase its biological activity [73, 74]. Therefore, in order to modify the emulsification activity of GMP, this peptide has been conjugated with other molecules such as lactose [73] and fatty acids [74]. The conjugation of GMP with lactose was carried out through the reaction of Maillard, and this conjugate showed a better emulsifying capacity without significantly reducing the solubility of GMP [73].

Currently, foams have many industrial uses of great importance in the production of beer, soaps, whipped cream, shaving cream, aerosols, etc. The formation of a foam requires the participation of a surfactant capable of diffusing to the air/water interface to lower the surface tension. GMP complies with this property, although the foams formed with GMP are stronger or more stable when combined with other foaming proteins [75, 76]. In order to improve the foam properties of the proteins, synergistic mixtures of biopolymers and pH variations have been made that can modify their charge and, consequently, their foam ability. In relation to this, by combining sodium caseinate with GMP, synergistic interactions take place between these molecules on foaming and on stability at pH 5.5 [77]. Non-glycosylated GMP has better foaming properties than glycosylated GMP [78]. This is due to the glycosidic structures favor a combination of hydrophilic and electrostatic effects, which prevents an orderly adsorption of the glycosylated GMP molecules at the air/water interface; whereas, non-glycosylated GMP forms a very stable network at the interface.

On the other hand, gels are semi-solid systems that consist of a network of solids (three-dimensional network of polymers) with an inside trapped-liquid. They are of great importance in food and pharmaceutical industry as many gelled products are manufactured throughout the world (gummies, gelatins, jelly jams, bakery fillings, and therapeutic or cleaning agents). Generally, gelling agents are proteins and polysaccharides. Gelling properties of GMP has been studied and it is known that its gelation depends on pH and temperature, reporting that even aqueous solutions with low GMP amounts can be gelled at pH below 4 [79]. Besides, GMP can potentiate gellying capacity of other substances. Thus, by fermenting goat milk to which GMP was added, a more ordered and structured gel was obtained, in addition to obtaining a better elasticity in it, as compared to that obtained when whey protein concentrate was added [80]. The influence of GMP on the gelation made by gelatin has also been studied and when these two compounds are mixed, lower concentration of both substances are need to get a gel as compared with the ones need when they are used separately [81]. This synergistic effect in gelation is very important in the food industry for the preparation of desserts and foods based on gels.

4.4 Oral care products

Dental caries is one of the chronic diseases that most often affect humans. Due to the anticariogenic and remineralization properties demonstrated to GMP and previously reviewed in biological activities section, nowadays GMP is being incorporated to some oral care products [15–17].

5. GMP as an indicator of milk adulteration with cheese whey

One of the problems presented by the dairy industry is the adulteration of milk with whey cheese, which is very cheap and not detected by sensorial tests. Cheese whey does not cause harm to health, however, it affects milk-derived products manufacturers financially and can affect the consumers nutritionally, so the addition of cheese whey is considered a fraud. Due to GMP present in cheese whey, the detection of this peptide may indicate the addition of cheese whey to milk. Some of the methods that detect GMP as an indicator of the presence of cheese whey are described below.

5.1 Electrophoretic methods

Sodium dodecyl sulfate–polyacrylamide gel (SDS-PAGE): This method was standardized to analyze pasteurized milk and milk powder. Samples has to be previously treated with 24% trichloroacetic acid (TCA) to eliminate interfering k-casein and later with 50% TCA to precipitate GMP, which is resuspended in Tris-buffer 0.05 M HCl, 1 mM EDTA, pH 7.2 [82]. Analyzing under these conditions cheese whey and milk added with cheese whey, a protein fraction of 20.8 kDa is identified corresponding to GMP that allowed to detect adulteration with whey up to 1%. This protein fraction is not detected in samples of acid whey or in raw milk. This technique had been previously used to analyze milk drinks that were distributed in schools as part of a program of the Brazilian government [83]. However, the method showed sensitivity to detect 5% of added whey, probably because they did not treat the sample with TCA. In later works, it was able to detect 40 and 20 μg of GMP in samples analyzed by electrophoresis in SDS-PAGE and in cellulose acetate strips, respectively, due to the use of thiobarbituric acid and malachite green dye reactions instead of Coomassie blue as developing agents [84].

Capillary electrophoresis (CE): A variant of electrophoresis is CE, a technique that has the advantage of allowing a rapid detection of GMP. This method has been used to identify GMP as an indicator of the presence of cheese whey in buttermilk powder and skim milk. An advantage of this method is that it is usually reproducible, repeatable and sensitive; however, the interpretation of the results is difficult [85].

5.2 Chromatographic method

High performance liquid chromatography (HPLC) has been widely used to identify GMP as indicative of milk adulteration with cheese whey. In order to carry out the analysis, it is necessary to pre-treat the samples with TCA to precipitate proteins that can interfere (k-casein) and to concentrate GMP [86]. Similarly, a rapid and sensitive HPLC method on a gel permeation column was developed to detect GMP to follow the hydrolysis of k-casein by chymosin in milk [87]. The only pretreatment given to samples was addition of TCA (final concentration 8%) to precipitate the interfering caseins and whey proteins. This method was widely used by several researchers to analyze different samples, such as skimmed milk powder [88]. Cation-exchange chromatography has also been used to detect GMP, previously removing caseins from whey samples by precipitation with HCl at pH 4.6, neutralizing with TCA at 2–8% and analyzing supernatants [89]. On the other hand, a Reversed-Phase HPLC (RP-HPLC) method was developed and validated to separate and quantify GMP and was demonstrated to be precise, sensitive, and reliable [90].

The determinations were performed in the linear range of 15–200 μg/mL and the detection limit was 2 μg/mL. The method was applied to the analysis of rennet and acid whey, whey protein concentrates produced by the dairy industry, and also for the detection of rennet whey in powdered milks.

The European Commission uses two methods to detect the presence of cheese whey in milk: a gel permeation chromatography and subsequently a RP-HPLC as a confirmatory test [91]. However, it has been shown that the sensitivity of this method is affected by the presence of acidified rennet whey, which makes it difficult to detect the addition of whey [92]. Besides, the HPLC methodology used to analyze compounds like GMP in dairy products usually includes extractions with solvents, sample's preparation require a lot of time and reactives, the equipment is very sophisticated and demands trained personal.

5.3 Spectroscopy methods

Spectroscopy has also been used to detect GMP. The medium infrared spectroscopy (MIR) was used to analyze milk powder in order to detect GMP as adulteration parameter. Although this method is fast, it is not widely used because derived spectra are not very easy to interpret, in addition to its high cost [93]. On the other hand, by liquid chromatography/electrospray coupled to mass spectrometry, milk products were analyzed and it was able to quantify GMP from concentration of 10 pmol, although the method was not used to detect milk adulteration [94].

5.4 Immunochemical methods

Immunoassays are analytical methods of great application in the food area, and have the advantages that they are quick, sensitive, and that the sample to be analyzed requires little or no treatment. Several immunochemical methods have been developed in order to identify and quantify GMP in milk. Firstly, it is necessary to produce antibodies against GMP and later, these antibodies can be used for the development of the different immunochemical methods that detect it. Some of these assays are described below:

Enzyme-linked immunosorbent assay (ELISA): It is an immunoassay widely used to analyze foods. It has the advantage that is simple, sensitive, and fast, in addition to being inexpensive. Two main ELISA assays has been developed to detect GMP in milk samples. An inhibition ELISA method was performed to detect bovine rennet whey solids in skim milk powders that presented a detection limit of 0.1% (w/w) and used enzyme-labeled monoclonal antibodies against bovine k-casein [95]. On the other hand, Chávez and co-workers [96] developed a sandwich ELISA using polyclonal antibodies against GMP, that showed a limit of detection of 0.047% (w/w).

Western blot assay: As ELISA, this technique is an immunoassay designed to detect proteins in complex samples and also has great specificity. Using the same polyclonal antibodies against GMP previously mentioned, Chávez and co-workers developed a western blot system to detect GMP [97]. When analyzing cheese whey, this antibody recognized three protein fractions of 20.1, 14, and 45 kDa. The detection limit of the test was 0.5% (v/v) to liquid cheese whey and 0.001% (w/w) to whey powder.

Immunochromatographic lateral-flow assay: The development of immunochromatographic systems for quality control is a relatively new field of research and has been applied to milk. There are commercial immunochromatographic sticks which contain monoclonal antibodies specific to GMP labeled with colloidal particles that present a limit detection of 4% (v/v) of milk whey. Using these immunosticks, it has been possible to identify GMP in different samples of commercial milks [98].

Besides, it has been developed an immunochromatographic lateral-flow test that used two specific anti-bovine κ-casein monoclonal antibodies, with a detection limit of 15 ng/ml of GMP and 1% (v/v) of cheese whey [99].

In summary, different techniques and methods have been developed and used to detect GMP as an index of adulteration of milk with cheese whey. Some of them can also be used to quantify GMP in food products. The aim of this area of research is to achieve one that bring together being cheap, fast, easy to develop, and to interpret the results, with high sensitivity and a limited sample processing. These characteristics will allow people to use them at the time and place of milk reception.

6. Conclusions

GMP possesses several nutritional and health promoting properties. Among them, it exerts important modulatory effects on the immune system that are beneficial in a number of different inflammatory conditions. GMP immune response mechanism of action might be mediated by increasing healthy intestinal microbiota, by inhibiting splenocyte proliferation, by promoting both local and systemic regulatory environment, and also by directly modulating immune cell functions. More research is needed to support these findings, as we cannot exclude a possible effect of products derived from GMP digestion on *in vivo* immunomodulatory activity. Besides, GMP is a peptide of promising industrial potential. It has a unique heat stability and solubility under acidic conditions that may suggest several uses in food. Studies on the functional properties of GMP may indicate new possible uses as a food additive. On the other hand, several methods have been developed and applied to detect GMP as an indicator of milk adulteration with cheese whey; however, none of them has been established as an international official method. Rapid, reliable, and inexpensive tests to detect GMP should be worked out to readily detect those cases of adulteration.

Acknowledgements

We appreciate the support given to the Autonomous University of Aguascalientes for the publication of this chapter.

Conflict of interest

Authors declare that there is no conflict of interest between the authors of the chapter entitled: "Glycomacropeptide: Biological activities and uses."

Author details

Juan Jauregui-Rincón*, Eva Salinas-Miralles, Norma Chávez-Vela
and Mariela Jiménez-Vargas
Autonomous University of Aguascalientes, México

*Address all correspondence to: jjaureg@correo.uaa.mx

References

[1] Hernández-Rojas M, Vélez-Ruiz JF. Suero de leche y su aplicación en la elaboración de alimentos funcionales. Temas Selectos de Ingeniería de Alimentos. 2014;**8**(2):13-22

[2] Universidad Tecnológica del Valle del Mezquital. Citrato de calcio a partir de residuos agroindustriales [Internet]. 2017. Available from: http://www.milenio.com/opinion/varios-autores/universidad-tecnologica-del-valle-del-mezquital/citrato-de-calcio-a-partir-de-residuos-agroindustriales [Accessed: 30-05-2018]

[3] Nishanthi M, Chandrapala J, Vasiljevic T. Compositional and structural properties of whey proteins of sweet, acid and salty whey concentrates and their respective spray dried powders. International Dairy Journal. 2017;**74**:49-56. DOI: 10.1016/j.idairyj.2017.01.002

[4] Moreno FJ, López-Fandiño R. Estructura y funcionalidad del glicomacropéptido bovino. In: Fontecha J, Recio I, editors. Funcionalidad de Componentes Lácteos. Elche: Universidad Miguel Hernández; 2009. pp. 223-249

[5] Tomczak-Tullio L, Lazzari-Karkle EN, Bileski-Cândido LM. Review: Isolation and Purification of Milk Whey Glycomacropeptide. Boletim do Centro de Pesquisa de Processamento de Alimentos in Curitiba. 2007;**25**(1):121-132

[6] Thomä-Worringer C, Sørensen J, López-Fandiño R. Review: Health effects and technological features of caseinomacropeptide. International Dairy Journal. 2006;**16**:1324-1333. DOI: 10.1016/j.idairyj.2006.06.012

[7] Neelima SR, Singh Rajput Y, Man B. Chemical and functional properties of glycomacropeptide (GMP) and its role in the detection of cheese whey adulteration in milk: A review. Dairy Science and Technology. 2013;**93**:21-43. DOI: 10.1007/s13594-012-0095-0

[8] Kawasaki Y, Isoda H, Tanioto M, Dosako S, Idota T, Ahiko K. Inhibition by lactoferrin and κ-casein glycomacropetide of binding of cholera toxin to its receptor. Bioscience, Biotechnology, and Biochemistry. 1992;**56**(2):195-198. DOI: 10.1371/journal.pone.0059253

[9] Isoda H, Kawasaki Y, Tanimoto M, Dosako S, Idota T. Patent: Use of compounds containing or binding sialic acid to neutralize bacterial toxins. European patent. Publication date: Sep 5, 1990; publication number: 0 385 118 A2

[10] Nakajima K, Tamura N, Kobayashi-Hattori K, Yoshida T, Hara-Kudo Y, Ikedo M, et al. Prevention of intestinal infection by glycomacropeptide. Bioscience, Biotechnology, and Biochemistry. 2005;**69**(12):2294-2301

[11] Brück WM, Kelleher SL, Gibson GR, Graverholt G, Lönnerdal BL. The efects of a-lactalbumin and glycomacropeptide on the association of CaCo-2 cells by enteropathogenic *Escherichia coli*, *Salmonella typhimurium* and *Shigella flexneri*. FEMS Microbiology Letters. 2006;**259**:158-162. DOI: 10.1111/j.1574-6968.2006.00268.x

[12] Hermes RG, Molist F, Pérez JF, Gómez de Segura A, Ywasaki M, Davin R, et al. Casein glycomacropeptide in the diet may reduce *Escherichia coli* attachment to the intestinal mucosa and increase the intestinal lactobacilli of early weaned piglets after an enterotoxigenic *E. coli* K88 challenge. The British Journal of Nutrition. 2013;**109**(6):1001-1012. DOI: 10.1017/S0007114512002978

[13] Feeney S, Tyan JT, Kilcoyne M, Joshi L, Hickey R. Glycomacropeptide reduces intestinal epithelial cell barrier dysfunction and adhesion of entero-hemorrhagic and entero-pathogenic *Escherichia coli* in vitro. Food. 2017;**6**:93. DOI: 10.3390/foods6110093

[14] Rong Y, Lu Z, Zhang H, Zhan L, Song D, Wang Y. Effects of casein glycomacropeptide supplementation on growth performance, intestinal morphology, intestinal barrier permeability and inflammatory responses in *Escherichia coli* K88 challenged piglets. Animal Nutrition. 2015;**1**(2):54-59. DOI: 10.1016/j.aninu.2015.05.006

[15] Nesser JR. Patent: Dental anti-plaque and anti-caries agent. Publication date: Feb 19, 1991; publication number: 4, 994, 441

[16] Schüpbach P, Neeser JR, Golliard M, Rouvet M, Guggenheim B. Incorporation of caseinoglycomacropeptide and caseinophosphopeptide into the salivary pellicle inhibits adherence of mutans streptococci. Journal of Dental Research. 1996;**75**(10):1779-1788. DOI: 10.5167/uzh-1649

[17] Zhang YP, Gaffar A. Patent: Fluoride free dental remineralization. Publication date: Mar 27, 2001; publication number: US 6,207,138 B1

[18] Shlygin GK, Vasilevskaya LS, Chernikov MP, Nikol'skaya GV. Antigastrin action of glycomaropeptide. Byulleten Eksperimental'noi Biologii i Meditsiny. 1971;**72**(12):9-13

[19] Vasilevskaia LS, Stan EI, Chernikov MP, Shlygin GK. Inhibiting action of glycomacropeptide on stomach secretion induced by various humoral stimulants. Voprosy Pitaniia. 1977;**4**:21-24

[20] Stan EI, Chernikov MP. Physiological activity of kappa-casein glycomacropeptide. Voprosy meditsinskoĭ khimii. 1979;**25**(3):348-352

[21] Chernikov MP, Stan EI, Vasilevskaia LS, Shlygin GK. Heterogeneity of the kappa-casein glycomacropeptide and physiological activity of its fractions. Voprosy Pitaniia. 1979;**5**:22-26

[22] Stan EY, Chernikoz EY. Formation of a peptide inhibitor of gastric secretion from rat milk proteins in vivo. Biulleten' Eksperimental'noĭ Biologii i Meditsiny. 1982;**94**(2):1087-1089

[23] Stan EI, Groĭsman SD, Krasil'shchikov KB, Chernikov MP. Effect of kappa-casein glycomacropeptide on gastrointestinal motility in dogs. Biulleten' Eksperimental'noĭ Biologii i Meditsiny. 1983;**96**(7):10-12

[24] Yvon M, Beucher S, Guilloteau P, Le Huerou-Luron I, Corring T. Effects of caseinomacropeptide (CMP) on digestion regulation. Reproduction, Nutrition, Development. 1994;**34**(6):527-537

[25] Beucher S, Levenez F, Yvon M, Corring T. Effects of gastric digestive products from casein on CCK release by intestinal cells in rat. The Journal of Nutritional Biochemistry. 1994;**5**(12):578-584. DOI: 10.1016/0955-2863(94)90012-4

[26] Pedersen NL, Nagain-Domaine C, Mahé S, Chariot J, Rozé C, Tomé D. Caseinomacropeptide specifically stimulates exocrine pancreatic secretion in the anesthetized rat. Peptides. 2000;**21**(10):1527-1535

[27] Gustafson DR, McMahon DJ, Morey J, Nan R. Appetite is not influenced by a unique milk peptide: Caseinomacropeptide (CMP). Appetite. 2001;**36**:157-163

[28] Keogh JB, Clifton P. The effect of meal replacements high in

glycomacropeptide on weight loss and markers of cardiovascular disease risk. The American Journal of Clinical Nutrition. 2008;**87**:1602-1605. DOI: 10.1093/ajcn/87.6.1602

[29] Keogh JB, Wooton BW, Taylor CM, Janakievski F, Desilva K, Clifton PM. Effect of glycomacropeptide fractions on cholecystokinin and food intake. British Journal of Nutrition. 2010;**104**:286-290. DOI: 10.1017/S0007114510000280

[30] György P, Kuhn R, Rose CS, Zilliken F. Bifidus factor. II. Its occurrence in milk from different species and in other natural products. Archives of Biochemistry and Biophysics. 1954;**48**(1):202-208. DOI: 10.1016/0003-9861(54)90324-0

[31] György P, Jeanloz RW, Nicolai H, Zilliken F. Undializable growth factors for *Lactobacillus bifidus* var. Pennsylvanicus. European Journal of Biochemistry. 1974;**43**:29-33. DOI: 10.1111/j.1432-1033.1974.tb03380.x

[32] Azuma N, Yamauchi K, Mitsuoka T. Bifidus growth-promoting activity of a glycmcropeptide derived from human K-casein. Agricultural and Biological Chemistry. 1984;**48**(8):2159-2162. DOI: 10.1080/00021369.1984.10866469

[33] Proulx M, Gauthier SF, Roy D. Effect of casein hydrolysates on the growth of bifidobacteria. Le Lait. 1992;**72**:393-404. DOI: 10.1051/lait:1992428

[34] Idota T, Kawakami H, Nakajima I. Growth-promoting effects of N-actylneuraminic acid-containing substances on bifidobacteria. Bioscience, Biotechnology, and Biochemistry. 1994;**58**:1720-1722. DOI: 10.1271/bbb.58.1720

[35] Janer C, Peláez C, Requena T. Caseinomacropeptide and whey protein concentrate enhance *Bifidobacterium lactis* growth in milk. Food Chemistry. 2004;**86**(2):263-267. DOI: 10.1016/j.foodchem.2003.09.034

[36] Robitaille G. Growth-promoting effects of caseinomacropeptide from cow and goat milk on probiotics. Journal of Dairy Research. 2013;**80**:58-63. DOI: 10.1017/S0022029912000660

[37] Ntemeri A, Chonchúir FN, O'Callagan TF, Stanton C, Ross RP, O'Toole PW. Glycomacropeptide sustains microbiota diversity and promotes specific taxa in an artificial colon model of elderly gut microbiota. Journal of Agricultural and Food Chemistry. 2017;**65**:1836-1846. DOI: 10.1021/acs.jafc.6b05434

[38] O'Riordan N, O'Callagan J, Buttò LF, Kilcoyne M, Joshi L, Hickey RM. Bovine glycomacropeptide promotes the growth of *Bifidobacterium longum* ssp. infntis anda modulates its gene expression. Journal of Dairy Science. 2018;**101**:1-12. DOI: 10.3168/jds.2018-14499

[39] Chen Q, Cao J, Jia Y, Liu X, Yan Y, Pang G. Modulation of mice fecal microbiota by administration of casein glycomacropeptide. Microbiology Research. 2012;**3**(1):e3. DOI: 10.4081/mr.2012.e3

[40] Ravussin Y, Koren O, Spor A, LeDuc C, Gutman R, Stombaugh J, et al. Responses of gut microbiota to diet composition and weight loss in lean and obese mice. Obesity. 2012;**20**(4):738-747. DOI: 10.1038/oby.2011.111

[41] Sawin EA, De Wolfe TJ, Aktas B, Stroup BM, Murali SG, Steele JL, et al. Glycomacropeptide is a prebiotic that reduces *Desulfovibrio* bacteria, increases cecal short-chain fatty acids, and is anti-inflmatory in mice. American Journal of Physiology. Gastrointestinal and Liver Physiology. 2015;**309**:G590-G601. DOI: 10.1152/ajpgi.00211.2015

[42] Otani H, Monnai M, Hosono A. Bovine k-casein as inhibitor of the proliferation of mouse splenocytes induced by lipopolysaccharide stimulation. ZB MED Nutrition, Environment, Agriculture. 1992;**47**:512-515

[43] Otani H, Hata I. Inhibition of proliferative responses of mouse spleen lymphocytes and rabbit Peyer's patch cells by bovine milk caseins and their digests. Journal of Dairy Research. 1995;**62**:339-348. DOI: 10.1080/09540109309354801

[44] Otani H, Monnai M. Inhibition of proliferative responses of mouse spleen lymphocytes by bovine milk k-casein digests. Food and Agricultural Immunology. 1993;**5**:219-229. DOI: 10.1080/09540109309354801

[45] Otani H, Monnai M, Kawasaki Y, Kawakami H, Tanimoto M. Inhibition of mitogen-induced proliferative responses of lymphocytes by bovine k-caseinoglycopeptides having different carbohydrate chains. Journal of Dairy Research. 1995;**62**:349-357. DOI: 10.1017/S0022029900031046

[46] Otani H, Monnai M. Induction of an interleuken-1 receptor antagonist-like component produced from mouse speen cells by bovine k-caseinoglycopeptide. Bioscience, Biotechnology, and Biochemistry. 1995;**59**:1166-1168. DOI: 10.1271/bbb.60.1017

[47] Monnai M, Otani H. Effect of bovine k-caseinoglycopeptide on secretion of interleukin-1 family cytokines by P388D1 cells, a line derived from mouse monocyte/macrophage. Milchwissenschaft. 1997;**52**:192-196

[48] Otani H, Horimoto Y, Monnai M. Suppression of interleukin-2 receptor expression on mouse CD4 + T Cells by bovine k-caseinoglycopeptide. Bioscience, Biotechnology, and Biochemistry. 1996;**60**:1017-1019. DOI: 10.1271/bbb.60.1017

[49] Mikkelsen TL, Bakman S, Sørensen ES, Barkholt V, Frøkiaer H. Sialic acid-containing milk proteins show differential immunomodulatory activities independent of sialic acid. Journal of Agricultural and Food Chemistry. 2005;**53**(20):7673-7680. DOI: 10.1021/jf050398o

[50] Requena P, González R, López-Posadas R, Abadía-Molina A, Suárez MD, Zarzuelo A, et al. The intestinal antiinflammatory agent glycomacropeptide has immunomodulatory actions on rat splenocytes. Biochemical Pharmacology. 2010;**79**(12):1797-1804. DOI: 10.1016/j.bcp.2010.02.008

[51] Requena P, Daddaoua A, Guadix E, Zarzuelo A, Suárez MD, Sánchez de Medina F, et al. Bovine glycomacropeptide induces cytokine production in human monocytes through the stimulation of the MAPK and the NF-kappaB signal transduction pathways. British Journal of Pharmacology. 2009;**157**(7):1232-1240. DOI: 10.1111/j.1476-5381.2009.00195.x

[52] Li EW, Mine Y. Immunoenhancing effects of bovine glycomacropeptide and its derivatives on the proliferative response and phagocytic activities of human macrophagelike cells, U937. Journal of Agricultural and Food Chemistry. 2004;**52**(9):2704-2708. DOI: 10.1021/jf0355102

[53] Yun SS, Sugita-Konishi Y, Kumagai S, Yamauchi K. Glycomacropeptide from cheese whey protein concentrate enhances IgA production by lipopolysaccharide-stimulated spleen cells. Animal Science Technology (Japan). 1996;**67**:458-462. DOI: 10.2508/chikusan.67.458

[54] Monnai M, Horimoto Y, Otani H. Immunomodificatory effect of dietary bovine k-caseinoglycopeptide on serum antibody levels and proliferative

responses of lymphocytes in mice. Milchwissenschaft. 1998;**53**(3):131-139

[55] Jiménez M, Chávez NA, Salinas E. Pretreatment with glycomacropeptide reduces allergen sensitization, alleviates immediate cutaneous hypersensitivity and protects from anaphylaxis. Clinical and Experimental Immunology. 2012;**170**(1):18-27. DOI: 10.1111/j.1365-2249.2012.04631.x

[56] Ye L, Chen Q, Li W, Yan Y, Zhao P, Pang G, et al. Effect of casein glycomacropeptide on phagocytic cells and intestinal mucosa immune cells in mice. Food Science. 2014;**35**(9):234-240. DOI: 10.1021/acs.jafc.5b04757

[57] Daddaoua A, Puerta V, Zarzuelo A, Suárez MD, Sánchez de Medina F, Martínez-Augustin O. Bovine glycomacropeptide is anti-inflammatory in rats with hapten-induced colitis. Journal of Nutrition. 2005;**135**(5): 1164-1170. DOI: 10.1093/jn/135.5.1164

[58] Requena P, Daddaoua A, Martínez-Plata E, González M, Zarzuelo A, Suárez MD, et al. Bovine glycomacropeptide ameliorates experimental rat ileitis by mechanisms involving downregulation of interleukin 17. British Journal of Pharmacology. 2008;**154**(4):825-832. DOI: 10.1038/bjp.2008.138

[59] Roldán NR, Jiménez M, Cervantes-García D, Marín E, Salinas E. Glycomacropeptide administration attenuates airway inflammation and remodeling associated to allergic asthma in rat. Inflammation Research. 2016;**65**(4):273-283. DOI: 10.1007/s00011-015-0913-y

[60] Muñoz FC, Cervantes MM, Cervantes-García D, Jiménez M, Ventura-Juárez J, Salinas E. Glycomacropeptide attenuates inflammation, pruritus, and Th2 response associated with atopic dermatitis induced by 2,4-dinitrochlorobenzene in rat. Journal of Immunological Research. 2017;**2017**:6935402. DOI: 10.1155/2017/6935402

[61] Jiménez M, Cervantes-García D, Muñoz YH, García A, Haro LM Jr, Salinas E. Novel mechanisms underlying the therapeutic effect of glycomacropeptide on allergy: Change in gut microbiota, upregulation of TGF-β, and inhibition of mast cells. International Archives of Allergy and Immunology. 2016;**171**(3-4):217-226. DOI: 10.1159/000453035

[62] Galli SJ, Tsai M, Piliponsky AM. The development of allergic inflammation. Nature. 2008;**454**(7203):445-454. DOI: 10.1038/nature07204

[63] Chen Q, Liang Y, Zhu C, Yan Y, Pang G. Effects of casein glycomacropeptide on the early development of primary colorectal cancer in rats. Food Science and Human Wellness. 2013;**2**:113-118. DOI: 10.1016/j.fshw.2013.09.001

[64] Gong J, Chen Q, Yan Y, Pang G. Effect of casein glycomacropeptide on subunit p65 of nuclear transcription factor-B in lipopolysaccharide-stimulated human colorectal tumor HT-29 cells. Food Science and Human Wellness. 2014;**3**:51-55. DOI: 10.1016/j.fshw.2014.04.001

[65] Takahashi N, Asakawa S, Dosako S, Tadashi I. Patent: Hypoallergenic nutritive composition. French patent. Publication date: Dec 24, 1992; publication number: 2677850

[66] Mikkelsen TL, Rasmussen E, Olsen A, Barkholt V, Frøkiær H. I nmunogenicity of K-casein and glycomacropeptide. Journal of Dairy Science. 2006;**89**(3):824-830

[67] Ney DM, Stroup BM, Clayton MK, Murali SG, Rice GM, Rhor F, et al. Glycomacropeptide for nutritional management of phenylketonuria: A randomized, controlled, crossover trial.

American Journal of Clinical Nutrition. 2016;**104**:334-345. DOI: 10.3945/ajcn.116.135293

[68] Kawaguchi T, Shiraishi K, Ito T, Suzuki K, Koreeda C, Ohtake T, et al. Branched-chain amino acids prevent hepatocarcinogenesis and prolong survival of patients with cirrhosis. Clinical Gastroenterology and Hepatology. 2014;**12**(6):1012-1018. DOI: 10.1016/j.cgh.2013.08.050

[69] Ney DM, Yen EC-L, Nelson DW, Solverson P. Patent: Glycomacropeptide to improve women's health. Publication date: Feb 22, 2018; publication number: US-2018/0050086 A1

[70] Chung S, Moughan P, Awati A, Morton H. The influence of whey protein and glycomacropeptide on satiety in adult humans. Physiology & Behavior. 2009;**96**(1):162-168. DOI: 10.1016/j.physbeh.2008.09.022

[71] Martin-Diana AB, Frías J, Fontecha J. Emulsifying properties of whey protein concentrate and caseinomacropeptide of cow, ewe and goat. Milchwissenschaft. 2005;**60**(4):363-366

[72] Chobert JM, Touati A, Bertrand-Harb C, Dalgalarrondo M, Nicolas M-G. Solubility and emulsifying properties of kappa casein and its caseinomacropeptide. Journal of Food Biochemistry. 1989;**13**(6):457-473. DOI: 10.1111/j.1745-4514.1989.tb00413.x

[73] Moreno FJ, López-Fandiño R, Olano A. Characterization and functional properties of lactosyl caseinomacropeptide conjugates. Journal of Agricultural and Food Chemistry. 2002;**50**(18):5179-5184. DOI: 10.1021/jf020118u

[74] Wong PYY, Nakamura S, Kitts DD. Functional and biological activities of casein glycomacropeptide as influenced by lyophilization with medium and long chain fatty acid. Food Chemistry. 2006;**97**(2):310-317. DOI: 10.1016/j.foodchem.2005.04.019

[75] Thomae-Worringer C, Siegert N, Kulozik U. Foaming properties of caseinomacropeptide– 1. Impact of concentration and interactions with whey proteins. Milchwissenschaft. 2007;**62**(3):249-252

[76] Martínez MJ, Carrera Sánchez C, Rodríguez Patino JM, Pilosof AMR. Interactions between β-lactoglobulin and casein glycomacropeptide on foaming. Colloids and Surfaces. 2012;**89**:234-241. DOI: 10.1016/j.colsurfb.2011.09.022

[77] Morales R, Martínez MJ, Pilosof AMR. Synergistic effect of casein glycomacropeptide on sodium caseinate foaming properties. Colloids and Surfaces B: Biointerfaces. 2017;**159**:501-508. DOI: org/10.1016/j.colsurfb.2017.08.017

[78] Kreuß M, Krause I, Kulozik U. Influence of glycosylation on foaming properties of bovine caseinomacropeptide. International Dairy Journal. 2009;**19**:715-720. DOI: 10.1016/j.idairyj.2009.06.012

[79] Farias ME, Martínez MJ, Pilosof AMR. Casein glycomacropeptide pH-dependent self-assembly and cold gelation. International Dairy Journal. 2010;**20**(2):79-88. DOI: 10.1016/j.idairyj.2009.09.002

[80] Martin-Diana AB, Pelaez C, Requena T. Rheological and structural properties of fermented goat's milk supplemented with caseinomacropeptide and whey protein concentrate. Milchwissenschaft. 2004;**59**(7):383-386

[81] Morales R, Martinez MJ, Pilosof AMR. Dynamics of gelation, textural and microstructural properties of gelatin gels in the presence of casein

glycomacropeptide. Food Research International. 2016;**84**:102-107. DOI: 10.1016/j.foodres.2016.03.024

[82] Galindo-Amaya L, Valbuena-Colmenares E, Rojas-Villarroel E. Standardization of glycomacropeptide detection with SDS–PAGE as a milk adulteration index. Revista Cientifica de Veterinaria (Maracaibo). 2006;**16**(3):308-314

[83] Souza EMT, Arruda FS, Brandão OP, Siqueira EM. Electrophoretic analysis to detect and quantify additional whey in milk and dairy beverages. Food Science and Technology. 2000;**20**(3):314-317. DOI: 10.1590/S0101-20612000000300007

[84] Nakano T, Ikawa N, Ozimek L. Detection of sialylated phosphorylated κ-casein glycomacropeptide electrophoresed on polyacrylamide gels and cellulose acetate strips by the thiobarbituric acid and malachite green dye reactions. Journal of Agricultural and Food Chemistry. 2007;**55**(7): 2714-2726. DOI: 10.1021/jf062987f

[85] Van Riel J, Olieman K. Determination of caseinomacropeptide with capillary zone electrophoresis and its application to the detection and estimation of rennet whey solids in milk and buttermilk powder. Electrophoresis. 1995;**16**(1):529-533. DOI: 10.1002/elps.1150160187

[86] Lieske B, Konrad G. A new method to estimate caseinomacropeptide and glycomacropeptide from trichloroacetic acid filtrates. Milchwissenschaft. 1996;**51**(8):431-436

[87] Van Hooydonk ACM, Olieman K. A rapid and sensitive high-performance liquid chromatography method of following the action of chymosin in milk. International Dairy Journal. 1982;**36**:153

[88] Olieman C, van den Bedem JW. A sensitive HPLC method of detecting and estimating rennet whey total solids in skim milk powder. International Dairy Journal. 1983;**37**(1-2):27-36

[89] Léonil J, Mollé D. A method for determination of macropeptide by cation-exchange fast protein liquid chromatography and its use for following the action of chymosin in milk. Journal of Dairy Research. 1991;**58**:321-328. DOI: 10.1017/S0022029900029897

[90] Ferreira I, Oliveira M. Determination of caseinomacropeptide by an RP-HPLC method and monitoring of the addition of rennet whey to powdered milk. Journal of Liquid Chromatography & Related Technologies. 2003;**26**(1):99-107. DOI: 10.1081/JLC-120017155

[91] Official Journal of the European Communities L 037. 2001. Regulation No. 213/2001, 07/02/2001. pp. 1-99

[92] Pádua É, D'Amico AL, Klaechim AJ, Walter EH, Garcia B, Fagnani R. Milk adulteration with acidified rennet whey: A limitation for caseinomacropeptide detection by high-performance liquid chromatography. Journal of the Science of Food and Agriculture. 2017;**98**(10): 3994-3996. DOI: 10.1002/jsfa.8846

[93] de Carvalho BMA, de Carvalho LM, dos Reis Coimbra JS, Minim LA, de Souza Barcellos E, da Silva Junior WF, et al. Rapid detection of whey in milk powder samples by spectrophotometric and multivariate calibration. Food Chemistry. 2015;**174**:1-7. DOI: 10.1016/j.foodchem.2014.11.003

[94] Mollé D, Léonil J. Quantitative determination of bovine k-casein macropeptide in dairy products by liquid chromatography/electrospray coupled to mass spectrometry (LC-ESI/MS) and liquid chromatography/electrospray coupled to tamdem mass spectrometry (LC-ESI/MS/MS). International Dairy Journal.

2005;**15**:419-428. DOI: 10.1016/j.idairyj.2004.08.013

[95] Bremer M, Kemmers-Voncken Boers EAM, Frankhuizen R, Haasnoot W. Enzyme-linked immunosorbent assay for the detection of bovine rennet whey powder in milk powder and buttermilk powder. International Dairy Journal. 2008;**18**:294-302. DOI: 10.1016/j.idairyj.2007.08.008

[96] Chávez NA, Jauregui J, Palomares LA, Macías KE, Jiménez M, Salinas E. A highly sensitive sandwich ELISA for the determination of glycomacropeptide to detect liquid whey in raw milk. Dairy Science and Technology. 2012;**92**(2):121-132. DOI: 10.1007/s13594-011-0052-3

[97] Chávez NA, Salinas E, Jauregui J, Palomares L, Macias K. Detection of bovine milk adulterated with cheese whey by western blot immunoassay. Food and Agricultural Immunology. 2012;**19**:265-272. DOI: 10.1080/09540100802381042

[98] Oancea S. Identification of glycomacropeptide as indicator of milk and dairy drinks adulteration with whey by immunocromatographic assay. Biotechnology Letters. 2009;**14**(1):4146-4151

[99] Martín-Hernández C, Muñoz M, Daury C, Weymuth H, Kemmers-Voncken AEM, Corbatón V, et al. Immunochromatographic lateral-flow test strip for the rapid detection of added bovine rennet whey in milk and milk powder. International Dairy Journal. 2009;**19**:205-208. DOI: 10.1016/j.idairyj.2008.10.016

Section 2

Buttermilk

Chapter 4

Technological and Biological Properties of Buttermilk: A Minireview

Andressa de Freitas Mascarello, Giovana Isabel Pinto, Isis Souza de Araújo, Letícia Kuller Caragnato, André Luís Lopes da Silva and Leandro Freire dos Santos

Abstract

Buttermilk is a by-product obtained from the batting insertion in the process of obtaining cream and butter and it is constituted by fat globules which are surrounded by milk fat globule membranes (MFGMs). During the stirring process, the membrane is ruptured and the various components present therein are released. Because it has a high nutritional content and low cost, buttermilk has drawn attention in the prospect of new forms of application. In addition, its disposal is expensive and not biologically viable. The objective of this work is to present a compilation of the technological and biological activities of buttermilk. Among the technological properties, it is worth mentioning its application as in the production of functional foods, a conduit for the incorporation of probiotics, inhibition of bacterial adherence on industrial surfaces, as well as the encapsulation of easily degraded activities and fermentative processes. Among the biological properties, its antioxidant, hypocholesterolemic, antimicrobial, and anticancer activities stand out. In conclusion, the reuse of buttermilk is economically and sustainably viable and encourages increasing research related to its use.

Keywords: buttermilk, technological properties, biological properties, butter, milk fat globule membranes

1. Introduction

From the batting inserted in the manufacturing process of the cream of fresh milk and butter, it is possible to obtain a by-product derived from the aqueous phase, called buttermilk [1, 2]. Buttermilk is made up of fat globules which are surrounded by membranes called milk fat globule membranes (MFGMs). Such membranes avoid its coalescence and its enzymatic degradation [1, 3]. During agitation, the fat globules are disrupted releasing various components dispersed into the aqueous medium such as polar lipids and fragments of MFGMs, which are possessing high functional potential due to their nutritional and technological characteristics [1, 2]. Exact quantity of production of buttermilk is not assessed; however, the quantity of production of buttermilk can be predicted on the basis of production of butter. Approximately 6.5–7.0% of total milk produced worldwide is used for the preparation of butter that yields high bulks of buttermilk as a by-product (around 3.2 million tons/annum) [4].

MFGM is rich in polar and neutral lipids, phospho- and sphingolipids, as well as mono-, di-, and triglycerides, as well as cholesterol [1]. Due to their technological and biological properties, the search has been increased by methods that concentrate and isolate MFGM [1, 3].

The use of buttermilk is justified due to its high nutritional value and low cost of obtaining it [5]. Its disposal in wastewater is uneconomical because it results in high BOD value (biochemical oxygen demand) [6]. Moreover, the lack of opportunities by not using this by-product in other industrial segments forces the dairy producers to continually invest in new technology for an environmentally correct disposal for this waste such as bioremediation [7].

On this front, the dairy industry is increasingly interested in using its by-products in new applications. The application of buttermilk is facilitated by the application of techniques such as pasteurization, concentration, and spray drying. Considering these aspects, the objective of the work was to gather studies that present the technological and biological properties of buttermilk.

2. Extraction, concentration, and analysis of phospholipids from the MFGM of buttermilk

Dairy glycerophospholipids, phosphatidylcholine, phosphatidylethanolamine, and phosphatidylinositol are the phospholipids commonly found in buttermilk. Of the class of sphingolipids, the main ones, found in buttermilk, are sphingomyelin, glucosylceramide, and lactosylceramide. The most commonly used phospholipid concentration processes are microfiltration (MF) and ultrafiltration (UF) [8]. The importance of extracting and concentrating these phospholipids is summarized by their bioactivity and their ability to act as emulsifiers.

MF has some limitations due to the size of casein fragments and MFGM phospholipids, which are very similar. Regarding the optimization of UF, studies performed pretreatments for the elimination of casein, predicting UF, precipitating it with acid, as well as the addition of agents that dissociate casein micelles such as citrate [8].

Isolation of MFGM has been hampered by the interaction of its components with other proteins during processing. The formation of the aggregates begins by ionic attraction between positively charged amino acid residues of the protein and the polar grouping of the lipids [9]. Researchers found that spraying, aiming at the drying of buttermilk, may induce the formation of MFGM protein and phospholipid complexes, making extraction of phospholipids difficult. It was further verified that in pasteurized buttermilk, there was an almost three times increase in the amount of MFGM-bound proteins compared to raw buttermilk. In addition, there was a high incorporation of β-lactoglobulin (β-LG) in MFGM isolated after heat treatment, and β-LG constitutes the major MFGM protein separated from pasteurized cream [10].

According to previous studies, the whey proteins of cheese may undergo thermal aggregation in the presence of concentrated buttermilk. In conclusion, its study showed that the phospholipids of the buttermilk membrane contributed to the formation of heat-induced aggregates with whey proteins [9]. Other analyzes have reported that most mechanisms of protein aggregation depend on nucleation and, as a consequence, the onset is done through the formation of an aggregation nucleus. These protein aggregates can be used as fat mimics in low-fat cheese or as yoghurt texture modifiers [11].

A study developed a gradient solvent system for high-performance liquid chromatography (HPLC) coupled to the ELSD detector (evaporative detector with light scattering) to separate the lipids present in buttermilk. The ELSD detector is sensitive only to the mass of the vaporized analyte, not being limited by solvent flow or ambient temperature, thus allowing better analysis time and adequate level of sensitivity. Among the standards used for analysis in HPLC were cholesterol, monostearin, diolein, phosphatidylcholine, phosphatidylinositol, phosphatidylserine, phosphatidylethanolamine, sphingomyelin, and other neutral lipids. In this study, separation with high reproducibility and accuracy of all classes of lipids present in buttermilk powder, including phospholipids, was observed, with no previous fractionation stage. The total concentration of polar lipids found in buttermilk was about 30%, the predominant phospholipid being phosphatidylcholine, followed by phosphatidine ethanolamine and phosphatidylserine, and with a lower concentration of sphingomyelin [12].

3. Technological properties of buttermilk

3.1 Biofilm formation

The incorporation, using different proportions of corn starch and buttermilk to obtain biopolymers, as well as the influence of temperature on the polymeric structure, is reported in the literature [13]. The study was justified by the application of the biopolymer in sustainable packaging, replacing polymers derived from petroleum, and reusing the by-product of the manufacture of butter. To obtain the biofilm, the author incorporated the phospholipids present in the milk derivative into polytetrafluoroethylene. The plates, containing the polymers, were dried during the minimum period of 24 hours in two different temperatures at 2% relative humidity. The thickness of the biofilms was monitored, which remained as a surface density of 55 g/m^2. The samples remained in desiccators under temperature control. The results indicated that in the polymers in which the buttermilk was incorporated, there was initially a separation of the phases, resulting from a certain incompatibility between buttermilk and starch, which significantly affected the modulus elasticity and tensile strength. The heating of the films, added with buttermilk, promoted a positive impact in relation to the resistance, which is justified by the gel formation that, during drying of the film, reduces the critical concentration for its formation. The permeability to water was not altered by the addition of up to 50% of buttermilk. Under heating, there was a decrease in the mass transport property due to the greater adhesion of the components of the polymer mesh. Finally, the rheological study classified the polymer as Newtonian, in addition to having antioxidant properties when subjected to heat treatment by the release of active peptides, and without antimicrobial activity under *Listeria innocua* [13]. In general, buttermilk, rich in proteins, has good film-forming ability by having plasticizers such as lactose.

3.2 Beverage production

A study investigated the use of buttermilk from fresh buffalo milk for the production of carbonated beverages flavored with fresh mango, orange, and pineapple fruit, varying the fruit juice concentration between 18, 20, 22, and 24% v/v of sugar between 8, 10, 12, and 14% m/v. The buttermilk used was 0.8% acidity, being first prefiltered to remove casein clots, and then filtered in millipore systems to obtain ultrafiltered buttermilk. Fresh fruit juices were also filtered to remove possible fibers. After the filtration processes, the buttermilk acquired low viscosity, since a large fraction of the proteins and lipids were removed from it. Among the analyzed flavors, the drink with 24% v/v of fruit juice flavored pineapple and 12% of sugar was the most accepted during sensory analysis, possibly due to the smaller amount of total solids in relation to the others. However, the higher the total solids content, the lower the solubilization of CO_2, which made the carbonation process difficult. The use of the fruit juice in the formulation helped to mask the astringent, sweet, and/or sour taste of buttermilk color, taste, aroma, and palate, as well as the overall appearance and acceptability of the product. The beverage produced had a higher concentration of proteins, vitamins, and minerals than market samples, as well as better physicochemical properties, and therefore had a better nutritional quality than the other samples analyzed [6].

3.3 Buttermilk as an additive in the composition of culture media for fermentative processes

Acidic bacteria (AB) are widely used in the production of fermented products. The use of buttermilk as a growth medium for probiotic lactobacilli has been reported. Probiotic foods are those that have probiotic bacteria in their formulation which, when given in sufficient quantities, promote health benefits for the user [5, 14]. The benefits are related to both the direct effect of probiotics and active metabolites produced during the fermentation process. These products should have between 106 and 107 CFU/mL of probiotics, being *Lactobacillus* and *Bifidobacterium* the most commonly used probiotics [14].

Although buttermilk presents a carbon source, there is a lack of nitrogen, which is necessary for the growth of AB. In this regard, an investigation used yeast extract as a source of nitrogen and obtained a good performance of the three strains tested in a medium containing buttermilk and yeast extract [15]. Another study, with a small concentration of yeast extract (0.3%), also perceived a good performance in the growth of *Lactobacillus* in buttermilk [16].

3.4 Application of buttermilk in the treatment of industrial surfaces

The procedures for controlling bacterial adhesion on industrial surfaces are of extreme importance, since a short time of contact with the surface is enough for the bacteria to begin biofilm formation. In this sense, buttermilk may aid in the inhibition of the formation of bacterial biofilms on industrial surfaces due to the high concentration of MFGM, which possesses polar lipids, which in turn affect the adhesion of bacteria on the industrial surface, preventing the formation of bacterial biofilms [17]. Studies indicated that buttermilk inhibited the adherence of microorganisms—*Lactococcus lactis, Leuconostoc cremoris, and Lactobacillus casei*—on stainless steel surfaces for 720 minutes, while other products, such as skim milk, were able to reduce bacterial adherence for about 30 minutes of exposure, which is considered a short time for the function [17].

3.5 Buttermilk encapsulation properties

A research examined the property of encapsulation of buttermilk in order to entrap omega-3 and thus be able to stabilize O/W emulsions. Omega-3 is a fatty acid with high demand due to its functional properties. A problem in their manufacture is their sensitivity to oxidation that hinders the delivery process to their place of absorption. Therefore, one of the strategies used to maintain its stability is encapsulation, causing its components to be confined within a matrix or within a small capsule, keeping them isolated from the external environment until, through an external stimulus, being released. In addition, buttermilk has emulsifying properties that aid in encapsulation. This emulsifying property is due to the presence of, for example, phospholipids and proteins, and other components such as liposomes that can be obtained from MFGM from buttermilk [18].

Buttermilk is still a good carrier of curcuminoid substances due to the presence of proteins and lipids in its composition, including phospholipids from MFGM. Curcuminoids are polyphenols from turmeric (*Curcuma longa*), which contain about 70–80% curcumin, 15–25% desmethoxycurcumin, and 3–10% bisdemetoxyc-curcumine, with anti-inflammatory, antioxidant, anti-HIV and as they are protectors against Alzheimer's disease, cystic fibrosis, and colon cancer [19]. A difficulty found in its clinical use is its low bioavailability, since it is poorly absorbed by the intestine because it has low aqueous solubility and low stability in pH near the neutral, present in the intestine, causing its hydrolysis to occur in smaller compounds [19, 20]. Curcumin has the ability to interact with MFGM lipids and proteins [19]. Some authors studied the influence of the use of buttermilk as a carrier for the curcuminoids. During storage of the curcuminoid with buttermilk and with buffer solution only, there was less degradation of the curcuminoids that were stored with buttermilk [19, 20]. This can be explained by the interaction of the same with the hydrophobic region of buttermilk that prevents its hydrolysis in aqueous environment. The encapsulation of actives has emerged as suitable vehicle for overcoming pharmacokinetic limitations associated with conventional drug formulations. Oftentimes, these features include incorporation of active targeting moieties for enhanced uptake in specific cells or constituent components for stimulus-responsive release (e.g., pH-sensitive, thermosensitive and ultrasound). Considering the contents discussed in topic (Buttermilk encapsulation properties).

Like curcuminoids, transresveratrol is a bioactive compound found mainly in strawberry, red grape, and wine, but it has limitations on its clinical application due to its low solubility and stability in aqueous environment. Among its benefits, antioxidant, cardioprotective, anticancer, and anti-inflammatory activities are mentioned. The solubility of resveratrol in aqueous phase is 13.6 μg/g in phosphate buffer (pH 7.4), considered low, and a higher solubility in oils, 179.8 ng/g. Therefore, the association of transresveratrol with lipids, such as those of MFGM, as well as proteins present in buttermilk, helps in increasing their bioavailability [21].

4. Biological properties

Among the biological properties found in buttermilk, some are as follows:

4.1 Antioxidant activity

Peptides released during the enzymatic or thermal hydrolysis of buttermilk proteins have known antioxidant properties. An investigation studied samples of biofilms, with antioxidant potential, produced from buttermilk. The results were expressed as trolox equivalent antioxidant capacity (TEAC), that is, the concentration of buttermilk in g/L that produces the same inhibition of a Trolox solution (analogous solution of vitamin E) at 1 mmol/L. When analyzing the antioxidant character of the biofilms, it was observed that the effect was only noticed in the samples that went through heating, which was justified by the release of the peptides during the heat treatment of the biofilms [13].

4.2 Hypocholesterolemic activity

The pathophysiology of coronary diseases is well established in the literature and is related, among other factors, to high cholesterol and low-density lipoprotein (LDL).

Considering the importance of the control of serum levels for greater clinical benefit, different studies have been looking for nonpharmacological measures in patients of low risk, among them the consumption of buttermilk [22, 23]. The hypocholesterolemic action of buttermilk was studied using different methodologies that demonstrated a reduction in total cholesterol and other variables with consumption of the dairy derivative in the short term [22–24]. The effect is explained by phospholipids present in buttermilk that may be responsible for the hypocholesterolemic effect. The effect occurs through an intervention in the pathway of cholesterol synthesis, as well as in the intestinal absorption of cholesterol, lowering their blood levels. A study conducted a double-blind, randomized, placebo-controlled study. Volunteers between 18 and 65 years and body mass index (BMI) of 35 kg/m^2 were submitted to two consecutive treatments of 4 weeks each, in random order. Participants should maintain their diet, medication, weight, alcohol consumption, and smoking habits normally, with the exception of the days before blood sampling for the exams. Vitamins and food supplementation products were banned from interfering with the research. The formulations tested and ingested by participants contained 22.5 g buttermilk or placebo, which should be mixed with 250 mL water and sucralose in a shaker provided to the participants. The placebo contained varied dairy ingredients; however, such ingredients do not have in their MFGM composition present in the milk fluid. The usual food intake was evaluated through a specific questionnaire for dairy foods in three moments: at the beginning, after the first treatment, and after the second treatment. After the blood draw procedure volunteers, tests of total cholesterol, triglycerides and serum concentrations of LDL and HDL, C-reactive protein, latosterol, β-sitosterol, campesterol and plasma levels of PCSK9, also called convertase 1, an enzyme that interacts directly with LDL receptors, decreasing its metabolism [25]. In conclusion, consumption of short-term buttermilk significantly reduced total serum cholesterol and triglyceride concentrations.

A manuscript evaluated 100 participants aged from 18 to 65 years regarding the lipid profile of individuals who ingested buttermilk. The study was randomized and placebo-controlled for a period of 12 weeks. The buttermilk drink was traditionally prepared with increments such as vanilla sugar in order to increase the acceptability of the dairy product. Participants were instructed not to change their usual diet, level of physical activity, and alcohol use. Total cholesterol, HDL, LDL, triglycerides, liver, and kidney function parameters, as well as cholesterol precursors, were quantified using HPLC. The author also noted a reduction in cholesterol and LDL levels [23].

4.3 Antimicrobial action

According to the World Health Organization (WHO), 30% of the population of industrialized countries in 2007 suffered from foodborne intestinal infections transmitted by food, creating, consequently, a public health problem. Due to the increased resistance of pathogenic organisms to antibiotics, new methods for gastrointestinal prevention and control are under investigation [26]. An investigation conducted an *in vivo* mouse study of the anti-infective effect of MFGM on *Salmonella enteritidis* and *Listeria monocytogenes,* a Gram-negative and a Gram-positive pathogen, respectively. The study demonstrated that ingestion of MFGM-rich buttermilk powder increased resistance to *L. monocytogenes* compared to skim milk with low amount of MFGM. This increase of resistance can be explained by the products of the phosphoglyceride and sphingolipid digestion, which showed to have antimicrobial activity *in vitro*. Another likely mechanism for resistance to *L. monocytogenes* colonization is that MFGM proteins promote inhibition of pathogen adhesion in the intestinal mucosa [26].

4.4 Anticancer action

Some authors studied the antiproliferative effect of several isolated fractions of buttermilk obtained from food and nonfood solvents on some types of human cancer cells. Fractions, rich in phospho- and sphingolipids, have strong antiproliferative action on colon and ovary cancer cells. These observations allow to hypothesize the use of these phospho- and sphingolipids as functional food additives [27].

5. Conclusion

A wide variety of applicability for buttermilk can be found in literature. Once obtained during the production process of the butter, it can be reused in different ways with technological or biological approaches. Its technological properties focus on the food and pharmaceutical industry as production of functional foods, a vehicle for the incorporation of probiotics, inhibition of bacteria adherence on industrial surfaces, as well as the encapsulation of easily degraded activities and fermentation processes. The biological properties focus on antioxidant, hypocholesterolemic, antimicrobial, and anticancer action. Thus, the wide variety of research presented, in different areas, stimulates new research related to this by-product.

Author details

Andressa de Freitas Mascarello[1], Giovana Isabel Pinto[1], Isis Souza de Araújo[1], Letícia Kuller Caragnato[1], André Luís Lopes da Silva[2] and Leandro Freire dos Santos[1*]

1 Department of Pharmacy, The Midwestern State University (UNICENTRO), Brazil

2 Department of Technology and Social Sciences, State University of Bahia (UEBA), Brazil

*Address all correspondence to: leandrofreire@onda.com.br

References

[1] Guggisberg D, Chollet M, Schreier K, Portmann R, Egger L. Effects of heat treatment of cream on the physical-chemical properties of model oil-in-buttermilk emulsions. International Dairy Journal. 2012;**26**(1):88-93

[2] De Bassi LG, Ferreira GCC, Da Silva AS, Sivieri K, Aragon-Alegro LC, Costa MDR. Evaluation of physicochemical, microbiological and sensorial characteristics of fermented milk beverages with buttermilk addition. International Journal of Dairy Technology. 2012;**65**(2):282-286

[3] Costa MR, Elias-Argote XE, Jimnez-Flores R, Gigante ML. Use of ultrafiltration and supercritical fluid extraction to obtain a whey buttermilk powder enriched in milk fat globule membrane phospholipids. International Dairy Journal. 2010;**20**(9):598-602

[4] Kumar R, Kaur M, Garsa AK, Shrivastava B, Reddy VP, Tyagi A. Natural and cultured buttermilk. In: Fermented Milk and Dairy Products. AK Puniya editor. USA: CRC Press, Taylor and Francis; 2015. pp. 203-225

[5] Burns P, Molinari F, Beccaria A, Páez R, Meinardi C, Reinheimer J, et al. Suitability of buttermilk for fermentation with *Lactobacillus helveticus* and production of a functional peptide-enriched powder by spray-drying. Journal of Applied Microbiology. 2010;**109**(4):1370-1378

[6] Shaikh MFB, Rathi SD. Utilization of buttermilk for the preparation of carbonated fruit-flavored beverages. International Journal of Dairy Technology. 2009;**62**(4):564-570

[7] Kosseva MR, Kent CA, Lloyd DR. Thermophilic bioremediation strategies for a dairy waste. Biochemical Engineering Journal. 2003;**15**(2):125-130

[8] Konrad G, Kleinschmidt T, Lorenz C. Ultrafiltration of whey buttermilk to obtain a phospholipid concentrate. International Dairy Journal. 2013;**30**(1):39-44

[9] Saffon M, Britten M, Pouliot Y. Thermal aggregation of whey proteins in the presence of buttermilk concentrate. Journal of Food Engineering. 2011;**103**(3):244-250

[10] Morin P, Jiménez-Flores R, Pouliot Y. Effect of processing on the composition and microstructure of buttermilk and its milk fat globule membranes. International Dairy Journal. 2007;**17**(10):1179-1187

[11] Saffon M, Jimnez-Flores R, Britten M, Pouliot Y. On the use of buttermilk components as aggregation nuclei during the heat-induced denaturation of whey proteins. Journal of Food Engineering. 2014;**132**:21-28

[12] Rodríguez-Alcalá LM, Fontecha J. Major lipid classes separation of buttermilk, and cows, goats and ewes milk by high performance liquid chromatography with an evaporative light scattering detector focused on the phospholipid fraction. Journal of Chromatography. A. 2010;**1217**:3063-3066

[13] Moreno O, Pastor C, Muller J, Atarés L, González C, Chiralt A. Physical and bioactive properties of corn starch–Buttermilk edible films. Journal of Food Engineering. 2014;**141**:27-36

[14] Antunes AEC, Silva ÉRA, Van Dender AGF, Marasca ETG, Moreno I, Faria EV, et al. Probiotic buttermilk-like fermented milk product development in a semiindustrial scale: Physicochemical, microbiological and sensory acceptability. International Journal of Dairy Technology. 2009;**62**(4):556-563

[15] Vigliengo E, Reinheimer J. Use of whey and buttermilk based media to obtain biomass of thermophilic LAB. International Journal of Dairy Technology. 2009;**62**(3):431-437

[16] Burns P, Vinderola G, Molinari F, Reinheimer J. Suitability of whey and buttermilk for the growth and frozen storage of probiotic lactobacilli. International Journal of Dairy Technology. 2008;**61**(2):156-164

[17] Dat NM, Manh LD, Hamanaka D, Van Hung D, Tanaka F, Uchino T. Surface conditioning of stainless steel coupons with skim milk, buttermilk, and butter serum solutions and its effect on bacterial adherence. Food Control. 2014;**42**:94-100

[18] Augustin MA, Bhail S, Cheng LJ, Shen Z, Oiseth S, Sanguansri L. Use of whole buttermilk for microencapsulation of omega-3 oils. Journal of Functional Foods. 2015;**19**:859-867

[19] Fu S, Shen Z, Ajlouni S, Ng K, Sanguansri L, Augustin MA. Interactions of buttermilk with curcuminoids. Food Chemistry. 2014;**149**:47-53

[20] Fu S, Augustin MA, Shen Z, Ng K, Sanguansri L, Ajlouni S. Bioaccessibility of curcuminoids in buttermilk in simulated gastrointestinal digestion models. Food Chemistry. 2015;**179**:52-59

[21] Ye J-H, Thomas E, Sanguansri L, Liang YR, Augustin MA. Interaction between whole buttermilk and resveratrol BT. Journal of Agricultural and Food Chemistry. 2013;**103**:7096-7101

[22] Conway V, Couture P, Richard C, Gauthier SF, Pouliot Y, Lamarche B. Impact of buttermilk consumption on plasma lipids and surrogate markers of cholesterol homeostasis in men and women. Nutrition, Metabolism, and Cardiovascular Diseases. 2013;**23**(12):1255-1262

[23] Baumgartner S, Kelly ER, van der Made S, Berendschot TTJM, Husche C, Lütjohann D, et al. The influence of consuming an egg or an egg-yolk buttermilk drink for 12 wk on serum lipids, inflammation, and liver function markers in human volunteers. Nutrition. 2013;**29**(10):1237-1244

[24] Severins N, Mensink RP, Plat J. Effects of lutein-enriched egg yolk in buttermilk or skimmed milk on serum lipids & lipoproteins of mildly hypercholesterolemic subjects. Nutrition, Metabolism, and Cardiovascular Diseases. 2015;**25**(2):210-217

[25] Corral P. Viewpoint back to basics: PCSK9 as a new target for the LDL receptor. Arquivos Brasileiros de Cardiologia. 2014;**102**:5-8

[26] Sprong RC, Hulstein MFE, Lambers TT, Van Der Meer R. Sweet buttermilk intake reduces colonization and translocation of *Listeria monocytogenes* in rats by inhibiting mucosal pathogen adherence. The British Journal of Nutrition. 2012;**108**:2026-2033

[27] Castro-gómez P, Rodríguez-alcalá LM, Monteiro KM, Ruiz ALTG, Carvalho JE, Fontecha J. Antiproliferative activity of buttermilk lipid fractions isolated using food grade and non-food grade solvents on human cancer cell lines. Food Chemistry. 2016;**212**:695-702